Lévy Statistics and Spin Glass Behavior in Random Lasers

Lévy Statistics and Spin Glass Behavior in Random Lasers

edited by
Anderson S. L. Gomes | André L. Moura
Cid B. de Araújo | Ernesto P. Raposo

JENNY STANFORD
PUBLISHING

Published by

Jenny Stanford Publishing Pte. Ltd.
101 Thomson Road
#06-01, United Square
Singapore 307591

Email: editorial@jennystanford.com
Web: www.jennystanford.com

British Library Cataloguing-in-Publication Data
A catalogue record for this book is available from the British Library.

Lévy Statistics and Spin Glass Behavior in Random Lasers

ISBN 978-981-4968-55-3 (Hardcover)
ISBN 978-1-003-33618-1 (eBook)

Contents

Foreword

It is customary for authors to ask an expert in a field to write a foreword for their book. Niels Bohr once said that "an expert is a person who has made all the mistakes that can be made in a very narrow field." I can say that in my case, I have made many mistakes in many fields and this is why I found this book to be fascinating, enlightening, and intellectually motivating.

The reasons that physicists can explain so much of nature is that we are able to perceive commonality between similar behavior in very different systems though the magical power we call physical intuition. The book *Lévy Statistics and Spin Glass Behavior in Random Lasers* exemplifies the way that nature can reveal itself and how we can increase the number of physical systems we "understand" through the realization of similarities and behavior tethered by our physical intuition.

During my earliest days in research, I was fortunate to be a part of the scientific awakening created by Edward Lorenz, a professor of meteorology at MIT. His work showed that the simplest set of equations describing a two-dimensional fluid layer uniformly warmed from below and cooled from above could result in periodic instabilities and chaotic behavior in a fully deterministic system. This discovery led to the generalized concepts of attractors and the sensitivity to initial conditions that revolutionized how we think about dynamical systems across many fields, including dynamos, electric circuits, chemical reactions, forward osmosis, and turbulence. With Haken's realization that the equations that govern the simplest of laser systems, a homogenously broadened, single-mode, and unidirectional laser, would be isomorphic to the Lorenz equations for the two-dimensional fluid, the hunt began for experimental validation. As my thesis was focused on CO_2

laser-pumped molecular lasers and they were in many ways an ideal system to observe the Lorenz–Haken instability, I became passionately involved with my graduate students in experimental and theoretical studies to elucidate the behavior of real systems and their underlying instabilities.

In 1994, Professor Anderson S. L. Gomes visited my labs at Brown University for what turned out to be a year of daily excitement. While the work on laser instabilities continued, my graduate student Mohan Balachandran, along with me, Professor Gomes, and Evelyne Sauvin, a postdoctoral researcher, began studying optical systems with very high gain and scattering. This led to the publication in the journal *Nature*, which launched the explosive interest in random lasers: "Laser Action in Strongly Scattering Media."

Light generation by stimulated emission in scattering systems was in fact not new and was first partially conceived by Ambartsumyan et al., who replaced a cavity mirror with a scattering material after which Letokhov predicted that it is possible that scattering alone can provide feedback for stimulated emission without any mirrors. The reason our work led to an explosive research interest in random lasers is that the scattering and amplification were decoupled and the system we devised could be easily made in any optics lab. This along with the extremely large number of high gain laser dyes spanning a large wavelength spectrum as well as the virtually unlimited number of scattering systems opened many experimental possibilities. In addition to the work aimed at elucidating the characteristics and theory of these systems, a huge number of potential applications emerged. The spectrum of applications included photomedicine, displays, search and rescue, combinatorial chemistry, sensors, and textiles. There were lasing paints, lasing clothing, and even lasing biological tissue.

The research work focused on both exploiting and elucidating the physics of random lasers was strongly intertwined with the research in light localization and photonic bandgaps. These areas of research led to the basic experimental and theoretical work which is a part of the current theories discussed in this book aimed at understanding the time evolution and emission spectrum of random laser systems.

Lévy Statistics and Spin Glass Behavior in Random Lasers provides, in a single book, both the experimental evidence and the current theoretical framework for understanding the behavior of random lasers with all of the key theoretical concepts clearly elucidated in each chapter. The book is an ideal resource for researchers who are interested in disordered systems and their connection to photonics. It exemplifies the power and value of finding analogies between physical systems and how the application of theoretical concepts from one field of physics or mathematics can expand our understanding of another.

The chapters in the book are ordered in a manner that takes the reader from the early concepts to the experimental data and establishes the connection to statistical mechanics of spin glasses as an isomorphic system. The application of Lévy statistics and the replica symmetry breaking in spin glasses is developed to explain random laser emission and is in many ways much like how Lorenz equations and geometrical attractors unraveled the seemingly unpredictable behavior of the most basic laser systems.

The first two chapters of the book provide a valuable review of the basic properties of random lasers of all shapes and forms, including fiber lasers and their general interconnection to statistical models and analogues such as spin glasses and turbulence. Chapter 1 focuses on establishing the underlying concepts of random lasers and also provides an authoritative review of the pioneering work on statistical physics, spin glass behavior, turbulence, mode correlation, and the Floquet phase of spin glasses. Chapter 2 focuses on the basic feedback mechanisms which can take place in a random system and the transition from the diffusive regime as originally proposed by Letokhov to the coherent near-localization regime where scattering is approaching the Ioffe–Regel limit of a mean free path equal to the central wavelength of the gain curve.

Chapters 3 and 4 leverage the basic concepts discussed in the first two chapters to elucidate the role of the Lévy flight distribution in random laser behavior with a very well-presented connection to experimental results (the reader would benefit from a study of associated passive systems which are purely driven by nonlinearities as well as scattering compositions which result in super-Gaussian light propagation). Chapter 4 discusses in greater detail the transition from Gaussian to Lévy statistics with heavily weighted tails and the approach of truncating the Lévy distribution function to avoid divergences from unbounded fluctuations.

In the second half of the book, Chapters 5, 6, 7, and 8 are focused on formulating and supporting the connection between random lasers and spin-glass systems. Chapter 5 begins with a review of the earliest works on the application of glassy phase transitions and replica symmetry breaking first developed for disordered magnetic spin systems (e.g., nonmagnetic metals doped with magnetic impurities and thiospinels such as indium diluted chromium $CdCr_{2x}In_{2(1-x)}S_4$) to random lasers. The concept of replica symmetry breaking, first applied to spin glasses to explain how different systems with the same long-range interactions (replicas) can result in distinct non-identical measurements, is applied to a multimode Hamiltonian formalism to make the connection to random lasers. Statistical methods are applied to obtain a phase diagram for the multimode random laser system with the analogy of inverse temperature in spin systems being analogous to the excitation or pumping rate in a random laser (as a point of interest, readers may want to investigate that the experiments on conventional solid-state liquid and dye lasers were also observed to exhibit the signatures of replica symmetry breaking near threshold in 2016). The final chapter carries the spin-glass physics to a specific case of a Floquet system. Such systems exhibit replica symmetry breaking in the presence of periodic coupling. This periodic coupling can be in either time or space, and in the case of random fiber lasers discussed in Chapter 8, it is temporal and originates from higher-order nonlinearities which drive pulsations.

I conclude by saying that this book is a valuable contribution to anyone interested in the current level of understanding of random laser systems and also in how it is that as physicists, we seem to understand so much. This book is a testament to the fact that nature can exhibit replicas in behavior across many physical systems.

Nabil M. Lawandy
Research Professor of Engineering
Brown University
Providence, Rhode Island

Preface

Scientific knowledge is nowadays disseminated at a breathtaking speed,

through conferences—presential, remote, or hybrid—scientific and even non-scientific journals and books. The building of such scientific knowledge in increasingly inter or multidisciplinary in nature, which requires the interactions among research scientists with different skills.

This book deals with a recent interdisciplinary approach, based on random lasers, encompassing theoretical and experimental studies in optics, statistical and complex physics, which grew over the last 25 years. The book was born due to the eyes and ears of visionary publishers, which after listening to a talk by one of the book editors (ASLG) in an international conference, invited to write the book. Being a multitask project, there are contributions from scientists who pioneered the field in pedagogically written chapters, what was first disseminated in peer-reviewed journals and conferences, such as to pave the way for other researchers to come up with new developments and achievements in the field. It also brings an updated state-of-the-art in this flourishing field.

The book is organized in 9 chapters, starting with a description in Chapter 1 of how the marriage among random lasers and random fiber lasers with spin glasses, Levy statistics, turbulence and Floquet states appeared in the literature, highlighting the main scientific aspects related to each topic. Chapter 2 brings a short introduction to the basics in random lasers and random fiber lasers, bringing attention to the feedback mechanisms and modes of random lasers, which are of paramount importance for the remaining of the book. In Chapter 3, analytical and numerical analyses of random lasers and Lévy statistics are reviewed, including the influence of high order nonlinear optical effects and extreme events, while in Chapter 4 the pioneer experimental work on Lévy statistics in random lasers is described. In Chapters 5 and 6, the authors report on the theoretical framework of spin glasses and its relation to photonics, and the first experimental works in the photonics analog are described. Chapter 7 brings an updated description of turbulence behavior in optics, with observations in fiber lasers and random fiber lasers. The phenomenon of Floquet spin-glass phase in random fiber lasers is described both theoretically and experimentally in Chapter 8. The book ends in Chapter 9, with a recent literature update on exploiting random lasers and random fiber lasers as photonic platforms for statistical physics and complex physics.

The endeavor of writing or editing a book is not a sole task, and several colleagues deserve some very wishful thanks. We start with our collaborators, who promptly accepted the task, Prof. Sushil Mujumdar, Prof. Neda Ghofraniha, Prof. Claudio Conti, as well as their co-workers. We also thank all involved MSc and PhD students, as well as post-docs from all the groups that contributed to the research described here. Without them, much of what was done would at least take more years to be achieved. Thanks also goes to our institution's staff, for all the help throughout our research projects.

We thank our families for the support and time we take from them, which make it easier on our work. Finally, we thank our financial sponsors, including CNPq, CAPES, and FACEPE, Brazilian Agencies, without which the research performed would not have been accomplished.

Anderson S. L. Gomes, André L. Moura,
Cid B. de Araújo, and Ernesto P. Raposo
Recife, PE, Brazil, 2022

Random Lasers and the Nobel Prize in Physics 2021

The Nobel Prize in Physics 2021 was awarded to Syukuro Manabe, Klaus Hasselmann and Giorgio Parisi "for groundbreaking contributions to our understanding of complex systems." In particular, Giorgio Parisi was rewarded "for the discovery of the interplay of disorder and fluctuations in physical systems from atomic to planetary scales."

The timing could not be more perfect as the Nobel Prize was announced (The Nobel Prize in Physics 2021. NobelPrize.org. Nobel Prize Outreach AB 2022. Sun. 11 Sep 2022. <https://www.nobelprize.org/prizes/physics/2021/summary/) just when we were finishing the last details of this book.

The work of Giorgio Parisi, from Università di Roma "La Sapienza," on complex systems in general and magnetic spin glasses in particular, is a milestone that later allowed the development of a whole new branch in the photonics of disordered nonlinear optical systems. Indeed, several parts of this book (e.g., Chapters 1 and 5–8) are devoted to exploring, both in the theoretical and experimental views, the so-called glassy behavior of light, which means that light can also display a photonic spin glass phase with replica symmetry breaking (RSB) much alike their magnetic counterpart studied by Parisi from the late 1970s.

The document "Scientific Background on the Nobel Prize in Physics 2021" (www.nobelprize.org/uploads/2021/10/sciback_fy_en_21.pdf) released by the Nobel Committee for Physics details the history of the spin glasses, from the early days of experimental discovery in disordered magnetic alloys to the first theoretical models and the seminal RSB solution by Parisi in 1979, and to the present day with applications to photonic random lasers and other systems.

In the early 1970s, a novel type of magnetic behavior quite different from the usual ones was observed in metallic alloys in which atoms of gold and iron were mixed in random positions. At very low temperatures, the spins of iron atoms point at random and fixed directions, as if they were "frozen." This new magnetic behavior was called "spin glass," in analogy with the randomness of molecules positions in a structural glass.

From 1975 onwards, the first theoretical models were proposed to explain this behavior. However, J. R. L. de Almeida and 2016 Nobel Prize winner in Physics, D. J. Thouless, found their spin glass solution to be unstable. Starting in 1979, Giorgio Parisi published a series of groundbreaking works that explained the behavior of spin glasses. According to this theory, two replicas of the system with identical thermodynamic behavior at high temperatures could present different properties in the spin glass phase at low temperatures. This phenomenon became known as "replica symmetry breaking" and gave rise to a real revolution in the study of disordered magnetic systems.

Almost thirty years after Parisi's seminal works, a group led by C. Conti, his colleague at the Università di Roma, theoretically predicted and experimentally demonstrated the glassy phase in another nature of systems, the photonic random lasers. Random lasers are quite different from conventional lasers, in which the laser beam is generated in a cavity with two mirrors. In random lasers, light is scattered by a large number of elements (e.g., atoms or ions) with random positions in the material. In this case, if certain conditions are satisfied, the emitted light has laser properties.

Conti and coworkers found that if the degree of randomness is high enough, then the phases of the electromagnetic waves that participate on the laser emission become correlated in a quite unusual way and "freeze" at random values, much like the freezing of spins in random directions in low-temperature magnetic spin glasses. This phase is now termed a photonic spin glass phase with replica symmetry breaking.

Quite shortly after these works, our group at the Universidade Federal de Pernambuco (UFPE) pursued further developments on this field, identifying, for example, the universality of the glassy transition in a variety of other random laser and random fiber laser systems (with distinct dimensionalities, structures, nature of random scatterers and gain medium, etc.). Since 2015, our research group has been also advancing the understanding of other complex behaviors in random lasers, including Lévy distributions of intensities, extreme statistical events, Floquet phases, and turbulent-like photonic phase.

Among the published work by the UFPE group, it has been highlighted by the Nobel Committee that one of the cited articles "links the early work of Hasselmann to that of Parisi and to the role of disorder and fluctuations in complex systems in general."

These advances have been approached in this book in an accessible level for beginners as well as for more experienced researchers. Many works associated with these developments have been cited in the mentioned document by the Nobel Committee for Physics.

Giorgio Parisi's Nobel Prize in Physics 2021 is definitively a landmark in the human endeavor to understand the ubiquitous complex systems. This book tells a bit of this history related to the photonic glassy phase and other remarkable features of random lasers and random fiber lasers.

Text prepared by the Book Editors

Chapter 1

Random Lasers, Lévy Statistics, Spin Glasses, Turbulence, and Floquet Phase: A Marriage between Photonics and Statistical/Complex Physics

Anderson S. L. Gomes, André L. Moura, Cid B. de Araújo, and Ernesto P. Raposo

Physics Department, Universidade Federal of Pernambuco,
Av. Prof. Luis Freire s/n, Cidade Universitaria,
Recife, PE 50670-901, Brazil
Anderson.lgomes@ufpe.br

1.1 Introduction

Photonics is defined as the science of generation, detection, and manipulation of light, whose quantum unit is the **photon** [1]. The aforementioned features have been experimentally accomplished through emission, transmission, modulation, signal processing, switching, amplification, and sensing of the light photons. Although most of the photonic applications are in the range of visible and near-infrared (400 nm to 2000 nm), several other applications extend to the mid- and far infrared up to 10,000 nm, or other regions

Lévy Statistics and Spin Glass Behavior in Random Lasers
Edited by Anderson S. L. Gomes, André L. Moura, Cid B. de Araújo, and Ernesto P. Raposo

ISBN 978-981-4968-55-3 (Hardcover), 978-1-003-33618-1 (eBook)
www.jennystanford.com

in the high frequency side of the electromagnetic spectrum, down to deep UV. The term photonics was coined in the 1960s and is a natural consequence of the spectacular growth and applications of the semiconductor lasers and LEDs invented in the early 1960s, and optical fibers in the 1970s. Together with erbium doped fiber amplifiers in the late 1980s, these three photonics components made it possible the boost in the telecommunications industry and its societal impact [2–5]. Applications of photonics are ubiquitous, and are part of our everyday life, with examples ranging from medicine (surgery, vision correction, endoscopy, health monitoring), laser material processing, the already mentioned telecommunications industry, information processing, lighting, holography, military technology, visual art, agriculture, and robotics, to the most advanced science, e.g. light detection, metrology, spectroscopy, biophotonics, nanophotonics, and many others.

For most of these applications, the advent of the LASER in 1960 [6] was of paramount importance. All technological applications and basic science development related to photonics was certainly made possible due to this fantastic source of photons, with its spatial, temporal and spectral coherence features [7]. The required ingredients for LASER operation are a pump source, a gain or amplifying medium and an optical cavity (a LASER resonator), usually composed of two mirrors (with one totally and the other partially reflector), known as a Fabry–Perot cavity [8]. A schematic of a conventional LASER is shown in Fig. 1.1.

Each LASER ingredient shown in Fig. 1.1 has its own task. The gain (or amplifying) medium can be a gas, liquid, solid, semiconductor, polymer in bulk or waveguide geometry form. Usually the "name" of the laser is given due to its gain medium. This is so since the first operation of the Ruby LASER, followed by a He-Ne, Ar, CO_2, Nd:YAG, GaAs, etc. The pump source can be optical, electrical, or even chemical. The LASER resonator can have different architectures, from the simplest Fabry–Perot cavity to the more sophisticated vertical cavity employed in semiconductor lasers [8] or fiber Bragg gratings (FBGs) employed in optical fiber lasers [9, 10]. Once the gain medium is properly excited by the pump source, electrons excited to high levels of the material's active atom can emit spontaneously after its spontaneous lifetime decay.

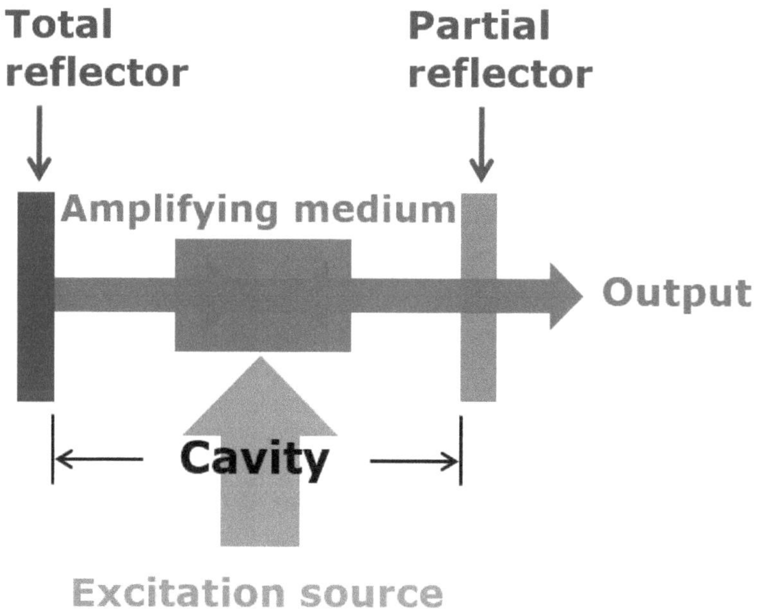

Figure 1.1 Conventional laser: the cavity, composed of two mirrors; the amplifying (or gain) medium, and (not shown) a source of input energy, which can be optical, electrical, or chemical.

If the levels in the atoms are appropriately chosen, population inversion can be achieved, leading to stimulated emission as predicted by Einstein in 1917 [11]. This stimulated emission can be amplified, made to oscillate in the resonator, and when the gain overcomes the losses inside the cavity (from scattering, mirrors, etc), coherent emission escapes the partially reflecting mirror in the form of a directional, narrowband—whose width is dictated by the combination of the cavity modes with the frequency where the gain bandwidth is highest—and temporally coherent. The type of source—continuous or pulsed—also determines whether the laser emission is continuous wave (cw) or pulsed. Pulsed sources can also be achieved from cw sources, by incorporating optical elements in the cavity, such as Q-switchers or mode-lockers, which can be active or passive devices. Tunability can also be achieved by incorporating specific dispersive elements in the cavity, such as gratings or prisms. These features of LASERS can all be reviewed in

Refs. [1, 7], or easily searched on the web. Nowadays, conventional LASERS are routinely used as "laser pointers", in many presentations and classroom lectures, or children's laser toys, which should be done with due care for safety [12] or to generate petawatt (PW, 10^{15} W) power in very short fs (10^{-15} s) pulses [13]. In between those extremes, two of the most widely applications of LASERS are in optical communications [14] and biophotonics [15]. Biophotonics includes basic science and applications in practically all medical branches: from ophthalmology to dermatology, including cancer therapy [16]. Conventional LASERS, as we qualitatively described above, are well-established photonics tools, and continue to be "a solution looking for a problem", as once said by unknown author. In the next section, we introduce a somewhat unconventional type of LASER which was named in 1995 a *random laser* (RL): It does not have the conventional mirror cavity, and the feedback mechanism responsible for laser oscillation is provided by scatterers. Following in the same steps, a *random fiber laser* (RFL) was developed in 2007. Still far from being utilized as much as the conventional LASER, both RLs and RFLs have already found interesting applications, therefore adding to the applications of conventional LASERS, as will be introduced in the next section.

1.2 Random Lasers and Random Fiber Lasers

In a 1966 publication by N. G. Basov's group, aiming at controlling the spatial coherence of a laser beam, Ambartsumyan and co-workers, among them Lethokov [17], worked on a different type of laser which they named "a laser with non-resonant feedback": The rear (100% reflecting) cavity mirror was replaced with a scattering surface. With this, they ensured that the light does not return to its original position after a round trip, so that the spatial resonances for the electromagnetic field are absent in the cavity. Central emission frequency of this laser was determined by the resonance frequency of a gain medium, instead of the eigenmodes of the cavity. Later, in a 1967 publication (in Russian) with English publication in 1968, V. S. Letokhov theoretically proposed the possibility of generating laser radiation in disordered dispersive media with gain [18]. The idea

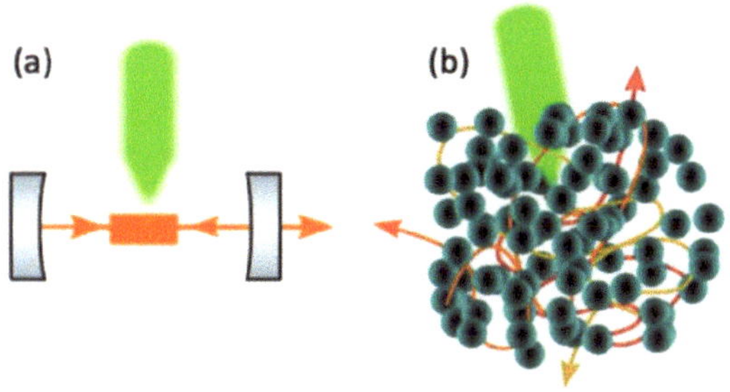

Figure 1.2 (a) Conventional laser scheme; (b) random laser idea: a gain media is embedded within the scatterers (represented by the balls), while it is pumped. The emission output is multidirectional.

was that radiation with the same characteristics of a conventional laser could be achieved in a dispersive and disordered medium *without the need for mirrors or resonant cavities.* Figure 1.2 shows a scheme comparing a conventional laser (2a) with the idealized random laser (2b).

The condition imposed by Letokhov [18] was that the mean free path of the photon should be much shorter than the dimension of the scattering medium. As reviewed in [19], several attempts to observe Lethokov's proposal were made after his publication, but it was only in 1994 that it was unambiguously demonstrated [20]. The seminal work of [20] triggered the important research area of RLs, as can be seen by the number of citation with the term Random Laser in the title in Fig. 1.3.

In optical fiber form, it is naturally known as RFLs, which has the advantage of inherent directionality, and was first demonstrated in 2007 [21]. It is nowadays a fast-growing field, as can be seen by the indicators in Fig 1.4.

Besides Ref. [19], several other review articles have been published describing the different RL systems, which include organic, polymeric, semiconductor, quantum dots, biomaterials, perovskites and optical fiber (doped and un-doped) as the gain medium, which can be optically or electrically pumped [22–31]. The authors call attention of the readers to the recent (2021) very broad review on recent advances and applications of RLs and RFLs of Ref. [32].

The most well-accepted definition of a random laser is "*A laser whose cavity is not formed by reflection from regular mirrors. Optical feedback comes from light scattering in a disordered medium.*" Figure 1.5 describes pictorially the typical types of gain media and architectures for RLs and RFLs.

Three main types of gain+scatterers combination can lead to RL emission, as identified in Fig. 1.5(a): The scatter is outside the gain medium (as exemplified by a colloidal RL); Fig. 1.5(b) the scatter and gain medium are the same, as exemplified by semiconductors (e.g. ZnO) or rare earth doped powders or optical fibers. Finally, Fig. 1.5(c) shows scatterers + gain micro/submicron or nanoparticles, which can combine metallic nanoparticles, leading to plasmonic enhancement, so-called plasmonic RLs [see Ref. 33 and refs. therein]. Most of the combinations shown in Fig. 1.5 can also be exploited in RFLs. It is also noticeable that several characteristics of conventional lasers that in principle would not appear to be present in RLs, have been overcome or mitigated. Those include tunability, unidirectionality, polarization control, Q-switching, and mode-locking [19, 32]. Very important to stress is that RLs and RFLs, although cavityless lasing devices, are not modeless. This and other characteristics of RLs will be treated in more detail in Chapter 2. Furthermore, it has been demonstrated that RLs follow Poissonian

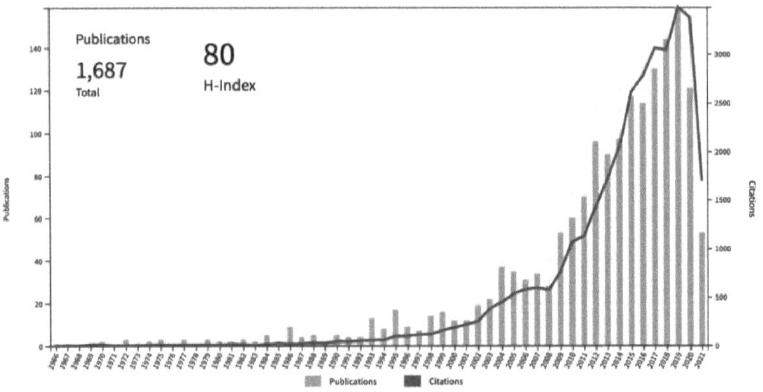

Figure 1.3 Number of citations per year, total of publications and h-index using the term Random Lasers in the title at the Web of Knowledge (downloaded July 13, 2021).

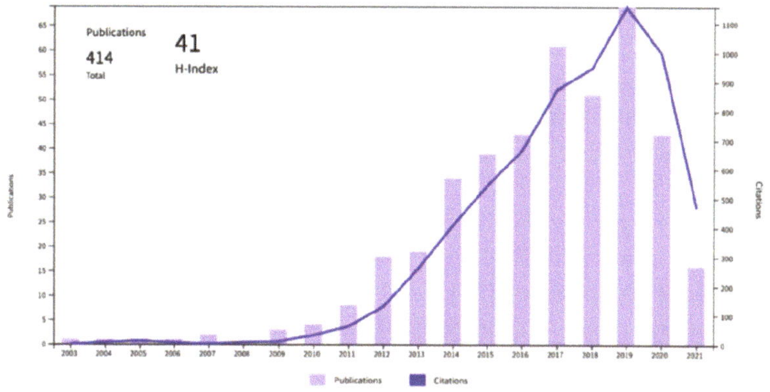

Figure 1.4 Number of citations per year, total of publications and h-index using the term Random Fiber Lasers in the title at the Web of Knowledge (downloaded July 13, 2021).

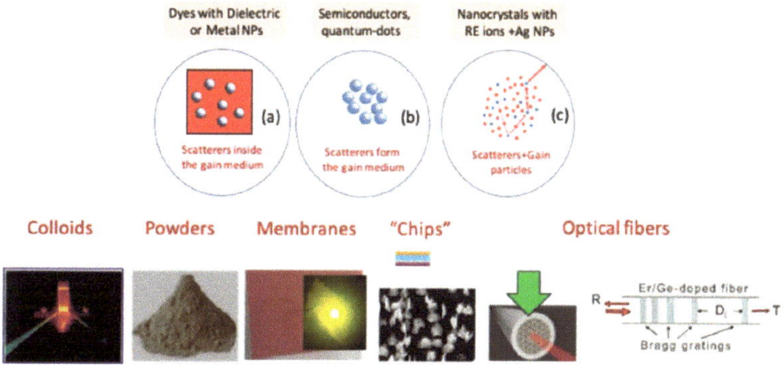

Figure 1.5 Combination of scattering and gain medium for random lasers. (a) Scatterers embedded in a gain medium (e.g., nanoparticles + dyes); (b) scatter is the gain medium (e.g., ZnO, rare earth ions); (c) gain particles + scatterers (e.g., rare earth ions + metallic nanoparticles).

statistics in the photon number, which is a characteristic of coherent optical sources [34]. Its low spatial coherence has led to applications as an excellent optical source for speckle-free images [35, 36]. Chapter 2 reviews the main aspects of the feedback mechanisms and modes of random lasers which are relevant to the subsequent chapters.

Random lasers are also intriguing because they are open, complex systems consisting of disordered scatterers embedded in gain media. These aspects have provided great possibilities for scientific insights into the broad field of complex systems in statistical physics and related phenomena, which is the main subject of this book. RLs have become a photonic platform for multi-disciplinary studies, as they are also a photonic platform for studies of astrophysical lasers, after the demonstration of random lasers in cold atoms [37]. In 2004, Anglos and co-authors [38] first experimentally reported that the intrinsic intensity fluctuations in RLs, not arising from the pump intensity fluctuation, deviated from Gaussian behavior, which triggered many theoretical and experimental studies of Lévy-like statistics in random lasers and random fiber lasers. Recent developments in this theme led to studies of the influence of high order nonlinear optical processes and extreme events. In another cross-disciplinary study exploiting RLs, it was proposed in 2006 by the Italian group led by C. Conti, in a theoretical article authored by Angelani and co-workers [39], that RLs would have glassy behavior, in analogy to spin glass phenomenon in statistical physics. This theoretical proposal was experimentally demonstrated by the same group in 2015 [40], again triggering considerable research. Yet another multi-disciplinary subject, turbulence, has been the object of intense research in quite different fields, from water turbulence to financial market [41], and has also been demonstrated in a random laser, in this case a random fiber laser [42]. An intriguing effect, also observed in a RFL [43] is the so-called Floquet phase, which is characterized by a time-dependent quantum Hamiltonian with a periodic driving source. In the remaining of this chapter, an introductory and historical overview of Lévy statistics, spin glasses, turbulence and Floquet phases will be given (next four sections) and the marriage between all of them with RLs and RFLs will close the chapter in Section 1.7.

1.3 Lévy Statistics

One of the major results of statistics theory is the so-called Central Limit Theorem (CLT) [44]. The incredibly wide range of applicability of the CLT in a rather broad variety of natural phenomena can be possibly attributed to its validity domain, which can be easily accessed provided only a few not uncommon requisites are fulfilled.

Consider, for instance, some phenomenon described by a given random variable. For example, the throw of a six-faces dice. Clearly, after a throw only one out of six results is possible, one for each face of the dice. However, consider now that the dice is thrown N times independently, and define the normalized sum variable $y = (x_1 + x_2 + \cdots + x_N)/N$, where x_i is the face shown after the i-th throw. In this case, it is clear that the random variable y can assume a huge number of possible values as N gets larger and larger.

The CLT states [44] that (i) if the random variables x_i are statistically independent and identically distributed (i.i.d.), (ii) if the second moment of the probability density function (PDF) of x_i is finite, and (iii) if N is sufficiently large, then the PDF of the sum variable y converges to a Gaussian distribution. In fact, this is the most common form of the CLT. Other variants do exist with slower or faster convergence to essentially the same Gaussian attractor, for instance, when the random variables are not identically distributed or when they keep some finite-range correlations, provided that some restrictions are obeyed [44].

One may thus argue what happens when some of the conditions above is not fulfilled. For example, consider i.i.d. random variables x_i whose PDF has *diverging* second moment. In this case, the CLT breaks down and is replaced by the so-called *generalized* CLT, whose statistical attractor, instead of the Gaussian, is the family of Lévy α-stable distributions [44]. These distributions were discovered in the 1930s by the French mathematician Paul Lévy. The main parameter that characterizes this statistical family is the Lévy index α, which assumes values in the range $0 < \alpha \leq 2$. Actually, the Gaussian distribution and CLT result are recovered for the boundary value $\alpha = 2$, which therefore justifies calling this theorem a *generalization* of the standard CLT.

In Chapters 3 and 4 of this book, we will give further details on the Lévy distributions. For the moment, it suffices to say that Lévy α-stable distributions are characterized by a diverging second moment and large fluctuations in the random variable. Perhaps due to this reason, for about six decades the use of Lévy distributions became somehow restricted to mathematical and statistical applications, with little contact with natural phenomena. Nevertheless, starting from the 1990s, a burst of phenomena in natural sciences emerged in the scientific literature which could not at all be suitably described by the Gaussian attractor and the CLT result. Instead, such novel phenomena were only possibly explained by probability distributions presenting very large second moment, which became immediately associated with Lévy, or at least Lévy-like, properties (a detailed discussion on this subtle matter will be presented in Chapters 3 and 4).

Among the most relevant examples of applications of Lévy statistics in the last two decades, we mention the distribution of values of output intensities emitted by RL and RFL systems, which will be considerably explored in the subsequent chapters as one of the main themes of this book. Extreme events and the influence of high order nonlinear optical processes in the intensity fluctuations in the Lévy regime of random lasers will also be addressed.

1.4 Spin Glass

Another central subject of this book is the emergence in the last decade of the photonic spin glass behavior in RL and RFL systems [39, 40, 45]. This will be the main theme of Chapters 5 and 6, but we anticipate in this chapter some introductory information.

We start by discussing magnetic systems without any kind of disorder. Consider, for instance, a ferromagnet in a three-dimensional crystalline structure with a large number of spins (conceivably, we comment below about the thermodynamic limit properties of the system). In high temperatures, the thermal fluctuations are so strong that no kind of long-range ordering sustains. In this case, the system is in the so-called paramagnetic phase, with spins pointing randomly along no preferential direction (a symmetric configuration invariant under rotations), with total magnetization null [46]. Importantly, in the high-temperature paramagnetic phase the dynamics of the system is very fast, so that spins are flipping and changing direction all the time. Now, as the temperature is lowered, a remarkable phenomenon takes place. Indeed, at some critical temperature, the rotational symmetry breaks down spontaneously and the spins start to align along a common direction. Such tendency continues as the temperature is further lowered, until all spins become fully aligned in the completely ordered ground state configuration. This characterizes the ferromagnetic state.

From the point of view of the evolution with temperature of the free energy functional of the system, in the paramagnetic phase only one global minimum of free energy is present, whereas in the ferromagnetic ground state two degenerate minima occur, one related to the configuration with all spins pointing, say, up, and the other associated with all spins pointing down. The paramagnetic-to-ferromagnetic transition is one of the most celebrated prototype examples of second order phase transitions in spatially ordered magnetic systems. Other kinds of magnetic phase transitions are also possible to spatially ordered configurations of spins, such as those leading to antiferromagnetic or ferrimagnetic states [46].

We shall now discuss what happens when some kind of disorder is introduced in the system. For example, consider that, instead of ferromagnetic couplings, the spins interact with random (positive and negative) strengths, in a form described, e.g., by a Gaussian distribution of couplings. Once again, for sufficiently high temperatures, the thermal fluctuations overcome the effect of the interactions between spins and the paramagnetic state sets in, with a single global minimum in the free energy landscape. However, for strong enough disorder (characterized, e.g., by a large enough variance in the Gaussian distribution of spin couplings) a phase transition can take place at a critical ("freezing") temperature to a rather distinct low-temperature state, defining the so-called spin glass phase [46].

In the spin glass state, a snapshot of the configuration of spins would, in principle, show a picture similar to that of the paramagnetic phase, with randomly pointing spins and null magnetization. Indeed, in the spin glass state the combined effect of disorder and magnetic frustration, due to conflicting interactions that each spin experiences, prevents the spins to align in a spatially ordered configuration. In this sense, it is commonly said that disorder and frustration are the main ingredients of magnetic spin glasses. However, the dynamics of the spin-glass and paramagnetic states are very different. Indeed, as the name suggests, in the spin glass phase the dynamics is very slow, in evident contrast with the paramagnetism phenomenon. Moreover, the free energy landscape is much distinct from both ferromagnetic and paramagnetic scenarios. In fact, in the spin glass phase the free energy functional develops a remarkable number of local minima that can eventually trap the system for a long time. This gives rise to metastability and irreversibility (e.g., hysteresis) in magnetic spin glasses.

From the theoretical point of view, a seminal model to describe spin glasses was proposed by Edwards and Anderson in 1975, a few years after the experimental discovery of the first spin glass in a magnetic alloy with random substitution of non-magnetic metals by a few percent of magnetic ions, reported by Cannela and Mydosh in 1972 [46]. Essentially, the Edwards–Anderson model considered a system of spins that interact only with nearest-neighbors via couplings randomly drawn from a Gaussian distribution [46]. Disorder and frustration elements are thus present in the Edwards–Anderson model, which was actually seminal to almost all subsequent theoretical attempts to explain the spin glass properties. For example, one of the two most important spin-glass models is the droplets model [46], in which the random interactions between spins are short ranged, as in the Edwards–Anderson model, so that the disorder and magnetic frustration make with that spins excitations occur in the form of clusters (or droplets) that, in order to evolve dynamically in time, must overcome energy barriers in the multivalley structure of the free energy landscape, whose height generally scales with a power-law of the cluster typical size.

The other seminal theoretical model for spins glasses is based on the Sherrington–Kirkpatrick Hamiltonian [46], which is essentially the infinite-range extension of the Edwards–Anderson model. In the mean-field approach to this Hamiltonian, all spins are considered to interact randomly with every other, with the same Gaussian distribution irrespective of the spatial distance between them. By employing the so-called "replica trick", in which identical copies of the system are coupled in the partition function formulation, the statistical physics solution of the Sherrington–Kirkpatrick model presented some anomalies in the specific heat and entropy at the ground state and, importantly, was later shown to be unstable precisely in the spin glass phase [46]. In an alternative approach, Parisi considered that the assumption of symmetry of replicas, perfectly valid in the paramagnetic state, could be broken in the spin glass phase [46]. This gave rise to the pioneering concept of replica symmetry breaking (RSB) in the spin glass phase, in which, e.g., systems identically prepared could yield different experimental measurements of observables, associated with the fact that they can be long trapped in distinct minima in the free energy landscape. Parisi's seminal idea influenced not only the physics of disordered magnetic systems, but also many branches of physics and even other scientific fields [46].

More recently, in a remarkable breath of renewal, spin glass ideas have also migrated to the field of photonics in the last decade. Indeed, as detailed in Chapters 5 and 6 of this book, starting in 2006 Angelani and collaborators developed [39] a photonic Hamiltonian analogy to the Sherrington–Kirkpatrick model that allowed the prediction and subsequent experimental confirmation of a spin-glass-like phase of light with RSB properties. In order to achieve this, the disordered photonic systems employed were the RLs and RFLs, as described in further detail ahead in this book.

1.5 Turbulence

Turbulence is a subject which—although its understanding has improved in the past decades—remains "the most important unsolved problem of classical physics" as Richard Feynman pointed out in the 1960s [47], and seems that it still remains as such, particularly quantum turbulence. Turbulent phenomena can be characterized by many time scales, which in turn can be related to the different scales of motion [48]. It can also be characterized by the existence of many degrees of freedom. As pointed out in [48], for instance, in fully developed hydrodynamical turbulence, these times are the turn-over times of eddies of different length scales. One important aspect of turbulence is the control parameter, whose increment is key for the phenomenon.

Turbulence can be found in several systems. Historically, the experimental studies of Osborne Reynolds (1883) on pipe flow led to introduction of the term turbulence by Lord Kelvin in 1887 [41]. Then statistical approach to describe the phenomenon was introduced in 1921 by G. I. Taylor. In 1941, A. N. Kolmogorov derived the so-called 2/3 and 4/5 laws [41]. Nowadays, turbulence is studied not only in natural hydrodynamical events, such as in water flows, atmospheric and oceanic environment, but also in engineered or man-made devices, as aircrafts engines, chemical reactions, etc. The nonlinear dynamics of chemical reaction can arise from internal or external interactions among species. In photonic systems, turbulence has been studied in fiber lasers, Bose–Einstein condensates and solitons, as reviewed in [42]. For a feeling of the complexity of turbulence phenomenon, atmospheric turbulence has shown to have an impact on optical communications systems, which can be mitigated by encoding of information in orbital angular momentum beams, in a context where Kolmogorov turbulence plays a relevant role [49].

Turbulence can be mathematically described in many ways, as reviewed in [41, 47], from the pioneer statistical methods developed by Kolmogorov to Navier–Stokes equations. These approaches can be complemented by other sets of partial differential equations which can describe turbulent systems as a manifestation of deterministic chaos. The statistical methods rely on the distribution function of the flow or the various moments of the velocity; in the dynamical methods, the mechanisms of instability and turbulent coherent structures are the primary focus. In Chapter 7, a theoretical description of turbulence-like phenomena applied to photonics and random fiber lasers is described, with a link to reported experiments. A Pearson-like correlation coefficient is introduced to simultaneously describe turbulence-like and photonic spin glass, as well as the use of Pearson correlation coefficient to study modes correlations in random lasers and random fiber lasers.

1.6 Floquet Spin-Glass Phase

The development and understanding of RLs and RFLs devices grew since 1994 [20] and 2007 [21], respectively, after the demonstration of their initial random lasing properties, and such devices can be classified as "novel" photon sources. Since then, the number of multidisciplinary applications has grown, as reviewed in Ref. [32], and quite interesting effects have already been studied exploiting the relationship between photonics and statistical physics or complex systems, as narrated in the former sections.

An intriguing effect occurs when the system is in non-equilibrium, which leads to the so-called *Floquet phases*. Unlike the photonic *equilibrium* spin-glass phase already introduced in this chapter, a Floquet phase arises from a *non-equilibrium* system, and its physical origin can be traced back to the periodic oscillations in time of the quadratic and quartic couplings in the system's photonic Hamiltonian, which, in turn, are related to a periodic oscillation of the source power. Such phenomenon has been first experimentally observed in a photonic system in a RFL [43], as will be thoroughly described in Chapter 8, and is a scientific problem that has several developments in the quantum world, and wherefore deserves further research.

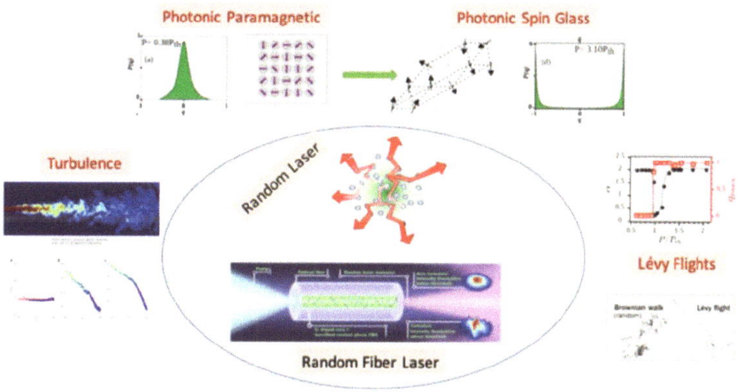

Figure 1.6 Pictorial description of the role of random lasers and random fiber lasers as a photonic platform for statistical and complex physics.

Table 1.1 Pioneer work on statistical physics and complex systems using random lasers and random fiber lasers, and their interconnections

Studied phenomenon	RL/RFL	Comments	References
Lévy-type behavior	In RL	First observation in ZnO nanoparticles in polymeric matrices did not point out Lévy behavior as the explanation for the intensity fluctuation [38], which was identified later [55, 56].	[38, 55, 56]
	In RFL	First observation in RFL exploiting Erbium ions as gain medium in a single mode fiber with randomly inscribed FBG as scatterers.	[57]
Spin-glass behavior	In RL	Theory and experiment (in 2D RLs) by the same group.	[39, 40]
	In RFL	First observation in RFL exploiting Erbium ions as gain medium in a single mode fiber with randomly inscribed FBG as scatterers.	[58]
Simultaneous Lévy + spin-glass	In RL	First simultaneous experimental observation in Nd^{3+} nanoparticles-based random lasers.	[50]
	In RFL	Observed in RFL exploiting Erbium ions as gain medium in a single mode fiber with randomly inscribed FBG as scatterers.	[57]
Turbulence-like	In RFL	Observed in RFL exploiting Erbium ions as gain medium in a single mode fiber with randomly inscribed FBG as scatterers.	[42]
Spin-glass+turbulence	In RFL	Observed in RFL exploiting Erbium ions as gain medium in a single mode fiber with randomly inscribed FBG as scatterers.	[52]
Pearson correlation/modes correlation	In RL	First exploited using nanofiber based RLs.	[59]
Simultaneous Pearson correlation + spin-glass	In RL	First demonstrated in Nd^{3+} nanoparticles-based random lasers.	[60]
	In RFL	First demonstrated in a hybrid RFL, exploiting Erbium doped fiber (EDF) and Semiconductor Optical Amplifier as a dual gain medium and Rayleigh scattering in the EDF.	[61]
Floquet phase	In RFL	Observed in RFL exploiting Erbium ions as gain medium in a single mode fiber with randomly inscribed FBG as scatterers.	[43]

1.7 The Marriage between Photonics and Statistical/Complex Physics

The marriage between RLs and RFLS with the Lévy statistics, spin glass, turbulence, and Floquet Phase were realized after 2016, when it was first noticed that Lévy-like and photonic spin-glass behavior could be observed from the same set of experimental data [50]. Thereafter, other publications have clearly shown, both theoretically and experimentally, that this is the case [51–54]. It is worth noting that other authors have demonstrated Lévy-like behavior, photonic spin-glass and turbulent—or more rigorously speaking turbulent-like—behavior, but the connection among them had not been identified. This will also be discussed in detail in this book. Furthermore, more recent work exploiting Pearson correlation in RLs and RFLs have been demonstrated and will also be described.

Figure 1.6 shows pictorially the role played by RLs and RFLs as photonic platforms for statistical and complex physics.

RLs and RFLs have been exploited for the observation of all above phenomena indicated in Fig. 1.6, by analyzing the statistics of the intensity fluctuations measured either in the spectral or temporal regimes. The pioneer works leading to this marriage are indicated in Table 1.1, highlighting the main characteristic of each study. In common to all of them is the possibility of using a photonic open cavity, with disorder and gain which surpasses losses and promote coherent (or low coherence) light emission, namely the RLs and RFLs, in either 3D, 2D, or 1D (sometimes quasi-1D) geometry.

References

1. Saleh, B. A. E., and Teich, M. C. (2019). *Fundamentals of Photonics*, 3rd ed., Wiley.
2. https://www.nap.edu/read/13491/chapter/1 (Optics and Photonics: Essential Technologies for Our Nation (2013).
3. https://www.iop.org/publications/iop/2018/file_71498.pdf (Downloaded May 5, 2019).

4. Girtan, M. (2014). Is photonics the new electronics? *Materials Today*, **17**, 100.
5. http://photonscanada.ca/media/55941/uk_photonics_revolutionising_our_world_jan_2014.pdf (downloaded May 5, 2019).
6. Maiman, T. H. (1960). Stimulated optical radiation in Ruby. *Nature*, **187**(4736), 493–494.
7. Siegman, A. E. (1986). *Lasers,* University Science Books.
8. Hodgson, N., and Weber, H. (2005). *Laser Resonators and Beam Propagation*, Springer Series in Optical Sciences.
9. Kashyap, R. (2009). *Fiber Bragg Gratings*, Academic Press.
10. Ter-Mikirtychev, V. (2014). *Fundamentals of Fiber Lasers and Fiber Amplifiers,* Springer Series in Optical Sciences.
11. Einstein, A. (1917). The quantum theory of radiation. *Physikalische Zeitschrift* **18**, 21.
12. https://www.fda.gov/media/86481/download (downloaded May 4, 2019).
13. https://www.laserfocusworld.com/articles/2017/02/llnl-hapls-petawatt-laser-reaches-highest-average-power-ready-for-delivery-to-eli-beamlines.html (downloaded May 4, 2019).
14. Klotzkin, D. J. (2014). *Introduction to Semiconductor Lasers for Optical Communications*, Springer.
15. Prasad, P. N. (2003). *Introduction to Biophotonics*, John Wiley & Sons, Inc.
16. Meglinski, I. (2014). *Biophotonics for Medical Applications*, Woodhead Publishing.
17. Ambartsumyan, R. V., Basov, N. G., Kryukov, P. G., and Letokhov, V. S. (1966). A laser with a nonresonant feedback. *IEEE Journal of Quantum Electronics*, **QE-2**, 442–446.
18. Letokhov, V. S. (1967). Generation of light by a scattering medium with negative resonance absorption. *Zhurnal Éksperimental'noĭ i Teoreticheskoĭ Fiziki* **53**, 1442–1447; *Soviet Physics JETP* **26** (1968), 835–840.
19. Luan, F., Gu, B., Gomes, A. S. L., Yong, K.-T., Wen, S., and Prasad, P. N. (2015). Lasing in nanocomposite random media. *Nano Today* **10**, 168–192.
20. Lawandy, N. M., Balachandran, R. M., Gomes, A. S. L., and Sauvain, E. (1994). Laser action in strongly scattering media. *Nature*, **368**(6470), 436–438.

21. de Matos, C. J. S., Menezes, L. D. S., Brito-Silva, A. M., Gamez, M. A. M., Gomes, A. S. L., and de Araújo, C. B. (2007). Random fiber laser. *Physical Review Letters,* **99**, 153903.

22. Sznitko, L., Mysliwiec, J., and Miniewicz A. (2015). The role of polymers in random lasing. *Journal of Polymer Science Part B, Polymer Physics,* **53**, 951–974.

23. Yu, S. F. (2015). Electrically pumped random lasers. *Journal of Physics D: Applied Physics,* **48**, 483001, 28.

24. Churkin, D. V., Sugavanam, S., Vatnik, I. D., Wang, Z., Podivilov, E. V., Babin, S. A., Rao, Y. J., and Turitsyn, S. K. (2015). Recent advances in fundamentals and applications of random fiber lasers. *Advances in Optics and Photonics,* **7**, 516–569.

25. Turitsyn, S. K., Babin, S. A., Churkin, D. V., Vatnik, I. D., Nikulin, M., and Podivilov, E. V. (2014). Random distributed feedback fibre lasers. *Physics Reports,* **542**, 133–193.

26. Wiersma, D. S. (2008). The physics and applications of random lasers. *Nature Physics,* **4**, 359–367.

27. Cao, H. (2005). Review on latest developments in random lasers with coherent feedback. *Journal of Physics A: Mathematical and General,* **38**, 10497–10535.

28. Cao, H. (2003). Lasing in random media. *Waves Random Media,* **13**, R1–R39.

29. Wiersma, D. S. (2013). Disordered photonics. *Nature Photonics,* **7**, 188—196.

30. Noginov, M. A. (2005). *Solid-State Random Lasers,* Springer, New York.

31. Kao, T. S., Hong, Y.-H. Hong, K.-B. and Lu, T.-C. (2021). Perovskite random lasers: a tunable coherent light source for emerging applications. *Nanotechnology,* **32**, 282001.

32. Gomes, A. S. L., Moura, A. L., de Araújo, C. B., and Raposo, E. P. (2021). Recent advances and applications of random lasers and random fiber lasers. *Progress in Quantum Electronics,* **78**, 100343.

33. Gomes, A. S. L. (2020). Nanocomposite-based random lasers: a review on basics and applications, in *Nanocomposites for Photonic and Electronic Applications,* Elsevier, pp. 45–79.

34. Cao, H., Ling, Y., Xu, J. Y., Cao, C. Q., and Kumar, P. (2001). Photon statistics of random lasers with resonant feedback. *Physical Review Letters,* **86**, 4524–4527.

35. Redding, B., Choma, M. A., and Cao, H. (2012). Speckle-free laser imaging using random laser illumination. *Nature Photonics,* **6**, 355—359.

36. Carvalho, M. T., Lotay, A. S., Kenny, F. M., Girkin, J. M., and Gomes, A. S. L. (2016). Random laser illumination: an ideal source for biomedical polarization imaging? *Multimodal Biomedical Imaging XI*, **9701**.
37. Baudouin, Q., Mercadier, N., Guarrera, V., Guerin, W., and Kaiser, R. (2013). A cold atom random laser. *Nature Physics,* **9**, 357–360.
38. Anglos, D., Stassinopoulos, A., Das, R. N., Zacharakis, G., Psyllaki, M., Jakubiak, R., Vaia, R. A., Giannelis, E. P., and Anastasiadis, S.H. (2004). Random laser action in organic-inorganic nanocomposites. *Journal of the Optical Society of America B, Optical Physics*, **21**, 208–213.
39. Angelani, L., Conti, C., Ruocco, G., and Zamponi, F. (2006). Glassy behavior of light. *Physical Review Letters,* 96, 065702.
40. Ghofraniha, N., Viola, I., Di Maria, F., Barbarella, G., Gigli, G., Leuzzi, L., and Conti, C. (2015). Experimental evidence of replica symmetry breaking in random lasers. *Nature Communications*, **6**, 7058.
41. Tsinober, A. (2009). *An Informal Conceptual Introduction to Turbulence*, Springer, 2nd ed.
42. Gonzalez, I. R. R., Lima, B. C., Pincheira, P. I. R., Brum, A. A., Macedo, A. M. S., Vasconcelos, G. L., de Menezes, L. S., Raposo, E. P., Gomes, A. S. L., and Kashyap, R. (2017). Turbulence hierarchy in a random fibre laser. *Nature Communications*, **8**, 15731.
43. Raposo, E. P., Gonzalez, I. R. R., Macedo, A. M. S., Lima, B. C., Kashyap, R., de Menezes, L. S., and Gomes, A. S. L. (2019). Evidence of Floquet phase in a photonic system. *Physical Review Letters*, **122**, 143903.
44. Uchaikin, V. V., and Zolotarev, V. M. (1999). *Chance and Stability: Stable Distributions and Their Applications,* VSP BV, Utrecht.
45. de Araújo, C. B., Gomes, A. S. L., and Raposo, E. P. (2017). (Review) Lévy statistics and glassy behavior of light in random fiber lasers. *Applied Science*, **7**, 644.
46. Mézard, M., Parisi, G., and Virasoro, M. A. (1987). *Spin Glass Theory and Beyond,* World Scientific, Singapore.
47. Feynman, R., Leighton, R. B., and Sands, M. (1964). *The Feynman Lectures on Physics*, Addison-Wesley.
48. Bohr, T., Jensen, M. H., Paladin, G., and Vulpiani, A. (1998). *Dynamical Systems Approach to Turbulence*, Cambridge University Press.
49. Malik, M., et al. (2012). Influence of atmospheric turbulence on optical communications using orbital angular momentum for encoding. *Optics Express,* **20**, 13195–13200.

50. Gomes, A. S. L., Raposo, E. P., Moura, A. L., Fewo, S. I., Pincheira, P. I. R., Jerez, V., Maia, L. J. Q., and de Araújo, C. B. (2016). Observation of Lévy distribution and replica symmetry breaking in random lasers from a single set of measurements. *Scientific Reports*, **6**, 27987, 8.
51. Raposo, E. P., Gomes, A. S. L., Moura, A. L., Fewo, S. I., Pincheira, P. I. R., Lima, B. C., Carreño, S. J. M., Silva, A. F., Maia, L. J. Q., Jerez, V., Gagné, M., Kashyap, R., and de Araújo, C. B. (2016). Universality in the Gaussian paramagnetic to Lévy spin-glass transition in random lasers. *Optics and Photonics News*, **27**, 34.
52. Gonzalez, I. R. R., Raposo, E. P., Macedo, A. M. S., de Menezes, L. S., and Gomes, A. S. L. (2018). Coexistence of turbulence-like and glassy behavior in a photonic system. *Scientific Reports*, **8**, 17046.
53. Uppu R., and Mujumdar, S. (20135). Exponentially tempered Lévy sums in random lasers. *Physical Review Letters*, **114**, 183903.
54. Gonzalez, I. R. R., Raposo, E. P., Macedo, A. M. S., Carreno, S. J., Maldonado, M., de Menezes, L. S., Gomes, A. S. L., and de Araújo, C. B. (2020). Influence of fifth-order nonlinear optical effects on the emission intensities statistics fluctuations in a photonic open-cavity complex system. *Physical Review A* **102**, 063515.
55. Sharma, D., Ramachandran, H., and Kumar, N. (2006). Lévy statistical fluctuations from a random amplifying medium. *Fluctuation and Noise Letters*, **6**, L95–L101.
56. Van der Molen, K. L., Mosk, A. P., and Lagendijk, A. (2006). Intrinsic intensity fluctuations in random lasers. *Physical Review A*, **74**, 053808.
57. Lima, B. C., Gomes, A. S. L., Pincheira, P. I. R., Moura, A. L., Gagné, M., Raposo, E. P., de Araújo, C. B., and Kashyap, R. (2017). Observation of Lévy statistics in one-dimensional erbium-based random fiber laser. *Journal of the Optical Society of America B Optical Physics*, **34**, 293–299.
58. Gomes, A. S. L., Lima, B. C., Pincheira, P. I. R., Moura, A. L., Gagné, M., Raposo, E. P., de Araújo, C. B., and Kashyap, R. (2016). Glassy behavior in a one-dimensional continuous-wave erbium-doped random fiber laser. *Physical Review A*, **94**, 011801-5(R).
59. Montinaro, M., Resta, V., Camposeo, A., Moffa, M., Morello, G., Persano, L., Kazlauskas, K., Jursenas, S., Tomkeviciene, A., Grazulevicius, J. V., and Pisignano, D. (2018). Diverse regimes of mode intensity correlation in nanofiber random lasers through nanoparticle doping. *ACS Photonics*, **5**(3), 1026–1033.

60. Coronel, E., da Silva-Neto, M. L., Moura, A. L., González, I. R. R., Pugina, R. S., Hilário, E. G., da Rocha, E. G., Caiut, J. M. A., Gomes, A. S. L., and Raposo, E. P. (2021). Simultaneous evaluation of intermittency effects, replica symmetry breaking and modes dynamics correlations in Nd:YAG random laser. Submitted to *Scientific Reports*.
61. Coronel, E., Das, A., González, I. R. R., Gomes, A. S. L., Margulis, W., von der Weid, J. P., and Raposo, E. P. (2021). Evaluation of Pearson correlation coefficient and Parisi parameter of replica symmetry breaking in a hybrid electronically addressable random fiber laser. *Optics Express*, **29**, 24422–24433.

Chapter 2

Feedback Mechanisms and Modes of Random Lasers

André L. Moura,[a,b] Pablo I. R. Pincheira,[c] Cid B. de Araújo,[d] and Anderson S. L. Gomes[d]

[a]*Campus Arapiraca, Universidade Federal de Alagoas, Arapiraca-AL, Brazil*
[b]*Programa de Pós-graduação em Física, Instituto de Física, Universidade Federal de Alagoas, Maceió-AL, Brazil*
[c]*Departamento de Ciencias Físicas, Universidad de La Frontera, Temuco, Chile*
[d]*Departamento de Física, Universidade Federal de Pernambuco, Recife-PE, Brazil*
andre.moura@fis.ufal.br

2.1 Basic of Lasers and Random Lasers

The basic ingredients of a conventional laser are the gain medium, an excitation source, and a cavity that can be due to a couple of mirrors. The laser operation requires a proper alignment of the mirrors, which are responsible for the feedback. In order to minimize the laser energy threshold, any scattering of light in the cavity should be avoided. On the other hand, random lasers (RLs) work based on the interplay between gain and scattering in disordered media with gain without an optical cavity to trap light: The scattering provides the feedback. To make a RL, one can grinding a laser crystal, and it

Lévy Statistics and Spin Glass Behavior in Random Lasers
Edited by Anderson S. L. Gomes, André L. Moura, Cid B. de Araújo, and Ernesto P. Raposo

ISBN 978-981-4968-55-3 (Hardcover), 978-1-003-33618-1 (eBook)
www.jennystanford.com

suffices to excite the powder with appropriate light [1]. Under these circumstances, the particles provide the gain and scattering. Other simple architecture for RL consists in ethanol solutions of laser dye and scatterer particles as TiO_2 [2], in which the gain is outside the scatterers. Several others systems have been investigated for RLs, like optical fibers [3], bones [4], cancerous tissue [5], etc.

2.2 Intensity Feedback: Photon Diffusion

RLs were proposed by Letokhov [6, 7] by investigating the generation of laser light in a disordered medium with gain ("negative absorption"). The study was carried out by analyzing the diffusion equation (Eq. 2.1) considering a photon density in a disordered medium with gain, neglecting the phases of the optical fields. In Eq. 2.1, $\Phi_\omega(\mathbf{r}, t)$ is the photon density at frequency ω, time t and position $\mathbf{r}$, D is the diffusion coefficient, c is the velocity of light, $Q_\omega(\mathbf{r}, t)$ the cross section for negative absorption (gain) at frequency ω, and N_0 the density of scatterer particles.

$$\frac{1}{c}\frac{\partial \Phi_\omega(\mathbf{r}, t)}{\partial t} = D\nabla^2\Phi_\omega(\mathbf{r}, t) + Q_\omega(\mathbf{r}, t)N_0\Phi_\omega(\mathbf{r}, t). \tag{2.1}$$

Since Eq. 2.1 is similar to the diffusion equation for monoenergetic neutrons in a homogeneous medium [8], Letokhov wrote the solution as

$$\Phi_\omega(\mathbf{r}, t) = \sum_n a_n\psi_n(\mathbf{r}) \exp\left[-\left(DB_n^2 - Q_\omega N_0\right) ct\right], \tag{2.2}$$

where $\psi_n(\mathbf{r})$ and B_n are the eigenfunctions and eigenvalues of Eq. 2.1 in the stationary regime, i.e., $\partial\Phi_\omega/\partial t = 0$, and a_n are arbitrary parameters determined by the initial conditions. A threshold condition is established from Eq. 2.2, since the photon density can decrease or increase with time. Then, the threshold is given by

$$DB^2 - Q_\omega N_0 = 0, \tag{2.3}$$

where B is the smallest eigenvalue among the many B_n values. As recognized by Letokhov, this threshold condition is analogous to the condition for critical multiplication of neutrons in a homogeneous nuclear reactor without a reflector. Letokhov visualized the particles working as scatterers and gain medium simultaneously, predicted the line narrowing due to the dependence of gain with the frequency, and because the only resonant element was the gain medium, he also proposed applications in highly stable optical frequency standard and the investigation of laser action in powders. Also, Letokhov predicted relaxation oscillations which are described ahead.

Many experimental attempts were made in order to demonstrate the theoretical proposal of Letokhov. In the 1970s, Fork et al. [9] observed unusual optical properties in microcrystals of different luminophosphors containing Eu^{2+}. The anomalously efficient and fast emission occurred in single microcrystals, the dimensions of which were on the order of several optical wavelengths $\lambda = 0.37\ \mu m$. In 1971, Varsanyi [10] experimentally observed optically excited stimulated emission ("super radiant emission") at the Pr^{3+} transition $^3P_0 \rightarrow {}^3F_2$ in individual powder particles of $PrCl_3$ and $PrBr_3$. The stimulated emission, which was not supported by any cavity, occurred in the crystalline grains with dimensions larger than 1 mm. In 1986 Markushev et al. [11] reported intense stimulated radiation from $Na_5La_{1-x}Nd_x(MoO_4)_4$ particle powders, and found out that above a certain excitation energy threshold, the duration of the emission pulse shortened by approximately four orders of magnitude. Later on, similar phenomenon was reported in a wide range of Nd^{3+} doped particle powders. However, in these earlier reports, the underlying physical mechanism was doubtful because it was known that a single micron-sized particle could support laser emission, which is known as whispery gallery modes [12]. That is different from Letokhov's idea, in which the diffusion of photons was the mechanism responsible for the optical feedback.

The unambiguous experimental demonstration of Letokhov's theoretical proposal was presented almost 25 years later. In 1994 Lawandy et al. [2] demonstrated laser-like emission from an optically pumped laser dye solution of Rhodamine 640 perchlorate in methanol. The light scatterer was due to particles of TiO_2 of average dimensions of ~250 nm. The authors found out that the emission from such systems can exhibit spectral and temporal properties of multimode laser oscillator, even though the gain medium was not contained inside an optical cavity, i.e., a spectral narrowing, and shortening of the temporal decay of the upper level

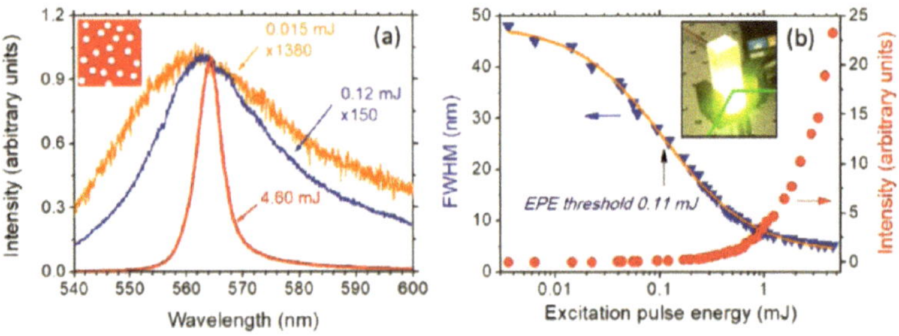

Figure 2.1 Example of characterization of dye based random lasers. (a) Output spectra for different excitation pulse energy. (b) Bandwidth and peak intensity as a function of the excitation pulse energy. The insets are a 2D representation of the colloid (a), and the colloid under large excitation energy exposure (b). Adapted from Ref. [14], with permission.

of the RL transition was observed. After this demonstration, the scientific community started to use the term Random Lasers (RLs) to describe the laser process in disordered materials with gain [13]. Figure 2.1 (from Ref. [14]) shows the usual characterization of a colloidal random laser: spectral narrowing (Fig. 2.1(a)), increasing in the output intensity when increasing the excitation pulse energy, and line width behavior versus the excitation pulse energy (Fig. 2.1(b)). A photograph of the operating RL is given in the inset of Fig. 2.1(b).

Recent approaches to the diffusion model consider the equations for the generated and the excitation photons, as well as and the atomic population, as in the work of Wiersma and Lagendijk [15]. A set of 3 coupled differential equations, which takes into account the energy density for the excitation light ($W_G(\mathbf{r}, t)$), the generated photons ($W_A(\mathbf{r}, t)$), and the concentration of laser particles ($N_1(\mathbf{r}, t)$), is written as in Eq. 2.4:

$$\begin{aligned}
\frac{\partial W_G(\mathbf{r}, t)}{\partial t} &= D\nabla^2 W_G(\mathbf{r}, t) - \sigma_{abs} \upsilon [N_t - N_1(\mathbf{r}, t)] W_G(\mathbf{r}, t) + \frac{1}{l} I_G(\mathbf{r}, t) \\
\frac{\partial W_A(\mathbf{r}, t)}{\partial t} &= D\nabla^2 W_A(\mathbf{r}, t) + \sigma_{em} \upsilon N_1(\mathbf{r}, t) W_A(\mathbf{r}, t) + \frac{1}{\tau_e} N_1(\mathbf{r}, t) \\
\frac{\partial N_1(\mathbf{r}, t)}{\partial t} &= \sigma_{abs} \upsilon [N_t - N_1(\mathbf{r}, t)] W_G(\mathbf{r}, t) \\
&\quad - \sigma_{em} \upsilon N_1(\mathbf{r}, t) W_A(\mathbf{r}, t) - \frac{1}{\tau_e} N_1(\mathbf{r}, t).
\end{aligned} \tag{2.4}$$

In these equations, υ is the velocity of light inside the random medium, σ_{abs} and σ_{em} are the absorption and emission cross sections, respectively, τ_e is the lifetime of the excited state, and N_t is the total concentration of laser particles. The diffusion constant (D) is given by $\upsilon l/3$, where l is the transport mean free path, i.e., the average distance that a wave propagates before its direction of propagation is randomized.

Noginov et al. [16] used a coupled rate equation system that was able to describe some RL characteristics, as increasing the input-output energy slope efficiency, and shortening of the time decay of the upper level of the RL transition. The model is described by coupled rate equations (Eq. 2.5) for the upper level of the RL transition, n, and the energy of the RL transition, E_{RL}. $K(t)$ is the excitation power density, whose temporal profile is considered to be Gaussian to mimic the excitation laser pulses. The parameter k_{abs} is the absorption coefficient of the Nd^{3+} at the excitation frequency, ν_{exc}, h is the Planck constant, τ_1 is the lifetime of the upper level of the RL transition in the absence of stimulated emission, c is the speed of light, ν_{RL} is the frequency associated with the excitation wavelength, σ_{RL} is the stimulated emission cross section, and τ_2 is the photon dwell time inside the powder. The factor $0 < \beta \leq 1.0$ is introduced to take into account that only a fraction of relaxing ions contributes to create photons at the RL transition [17].

$$\frac{dn}{dt} = \frac{K(t)k_{abs}}{h\nu_{exc}} - \frac{n}{\tau_1} - \frac{E_{RL}}{h\nu_{RL}}c\sigma_{RL}n$$
$$\frac{dE_{RL}}{dt} = -\frac{E_{RL}}{\tau_2} + \beta\frac{n}{\tau_1}h\nu_{RL} + E_{RL}c\sigma_{RL}n. \tag{2.5}$$

The main characteristics of RL emission, which are reproduced by Eq. 2.5, are the spectral narrowing (Fig. 2.2(a)), abrupt increase in the input-output energy slope efficiency (Fig. 2.2(b)), and shortening of the decay time of the upper level of the RL transition (Fig. 2.2(c)) when increasing the excitation intensity from below to above the RL energy threshold. This threshold is well defined in Fig. 2.2(b) by the abrupt bandwidth narrowing and increasing of the slope efficiency. It is worth noticing that in some RLs the transition is not abrupt and sometimes is confused with the amplification of the spontaneous emission [2, 14]. However, by analyzing intensity fluctuation correlations obtained under identical

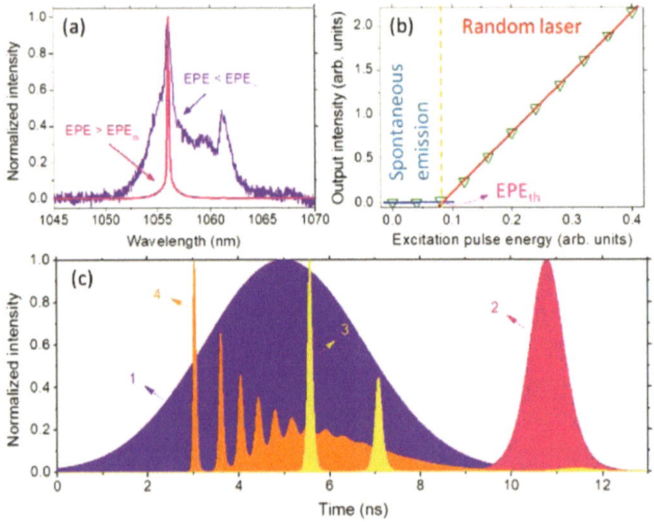

Figure 2.2 Random laser (RL) characterization for trivalent neodymium ions (Nd^{3+}) doped particles. (a) Experimental spectral narrowing observed when increasing the excitation pulse energy (EPE). Results from numerical simulation - (b) input-output energy curve. The RL EPE threshold is defined by the slope increase (c) Temporal characterization of the RL intensity behavior. Curve 1 corresponds to the excitation beam, which was considered a Gaussian with half width at half maximum of 4.1 ns centered at 5 ns. Curves 2, 3 and 4 are the RL emissions for EPE close, larger, and much larger than the threshold.

experimental conditions, it is possible to determine the transition from the spontaneous emission to the RL regime when increasing the gain [14, 18, 19] (see Chapter 6).

At this point, it is worth emphasizing that the wave nature of light was neglected in the models described above, i.e., interference effects and the presence of resonance modes are not considered.

2.3 Field Feedback: Interference Manifestations

Cao et al. [20, 21] demonstrated discrete laser lines in the RL spectra of ZnO nanoparticles powder. The mean free path of photons was on the order of the emission wavelength, and it was proposed

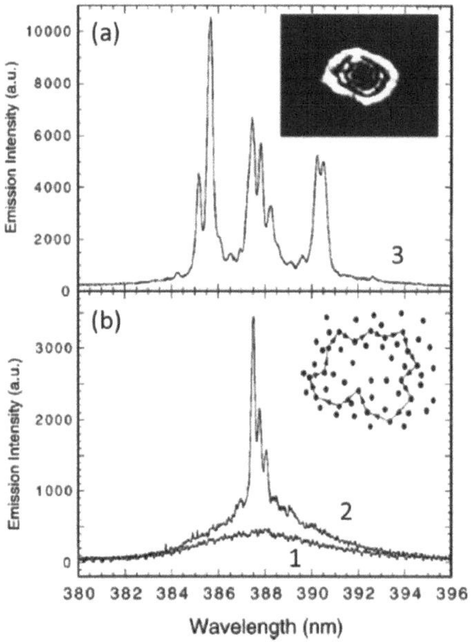

Figure 2.3 (a) and (b) Zinc oxide (ZnO) based random laser spectra for increasing excitation area (from bottom to top) 980 (curve 1), 1350 (curve 2), and 1870 μm^2 (curve 3). The excitation intensity was 1012 kW/cm^2. The inset in (a) corresponds to amplified images of the excitation area above the lasing threshold, and the inset in (b) is a representation of the closed paths for light. Adapted from Ref. [20] with permission.

that the emitted photons could make closed loop paths, as shown schematically in the inset of Fig. 2.3(a). The assumption of closed loops is because the mean free path of the photon within the medium is of the order of the light wavelength. Therefore, light can be "trapped" in the system forming ring-shaped cavities. Throughout the different loops there are different probabilities of having a resonance. The threshold for laser oscillation is defined when the amplification along a loop path exceeds the losses, and the associated wavelength is determined by requiring the phase shift along the loop equals to a multiple of 2π.

After the experimental evidence of recurrent light scattering in RLs [20–22], the literature divided the RLs in two types [23]: one with non-resonant (incoherent) feedback; and the other

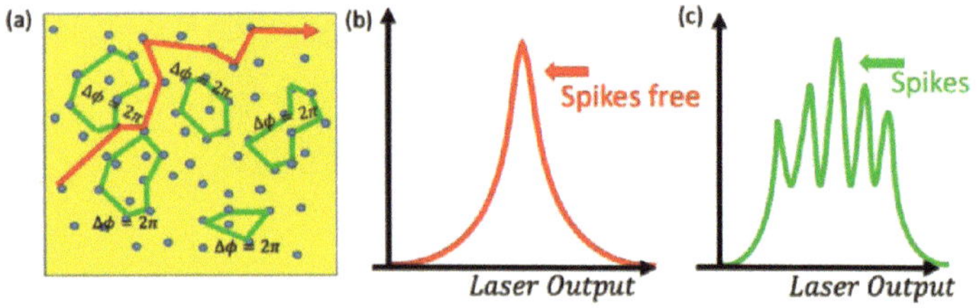

Figure 2.4 The random laser regimes are illustrated in (a), in which the red arrow shows the case of incoherent feedback while the green arrows represent the closed loops for the case of coherent feedback. (b) Smooth output spectrum (spikeless) corresponding to incoherent feedback. (c) Representative spectrum with spikes structures associated with the closed loop paths (a)—coherent feedback.

with resonant (coherent) feedback. In the former, one has only an intensity feedback, which means that interference effects are irrelevant, and are associated to open paths in the diffusive medium (Fig. 2.4(a)). The output spectra present a narrowing behavior when increasing the gain (Fig. 2.4(b)), and most of the experimental observations are described by the diffusion models in Section 2.2. On the other hand, for coherent feedback, the phases of the optical fields play the role, and the interference effect is unveiled by the presence of resonant wavelengths in the RL output spectra (Fig. 2.4(c)). The corresponding resonant modes are associate to a closed loop picture that emulates an optical mirror cavity. Figure 2.4(a) shows the conceptual idea of loops with phase shift of 2π (green arrow). The observation of one or the other regime depends on some characteristic lengths of the medium as scattering mean free path, and the size of the random medium, as described in the next section. Due to the similarities between the optical propagation through disordered systems and electron scattering in random potentials, Anderson localization, which is a manifestation of interference effects for electrons providing absence of diffusion [24], was introduced in the RL literature [25–29]. The localized regime is expected when the scattering mean free path is small enough to provide recurrent scattering. An equivalence between the localized modes in a RL and the cavity modes of a conventional laser was proposed [28].

2.4 The Regimes of the Random Laser and Its Characteristic Lengths

When light propagates through a scattering amplifying medium, the multiple scattering increases the path length or, equivalently, the residence time of the light in the active medium. Due to the inverted atomic population and the enlarged light paths, one has the amplification of light by stimulated emission. In analogy with a conventional laser, where the distance between the mirrors of the optical resonator is essential to determine the laser modes, in a RL there are relevant length scales which quantify the scattering process.

2.4.1 Scattering Mean Free Path, l_s

The scattering mean free path, l_s is the average distance that light travels between two consecutive scattering events and defined as

$$l_s = \frac{1}{n_s \sigma_s}, \tag{2.6}$$

where n_s is the number density of scatterers. σ_s is the scattering cross section of the individual scatterer and depends on three variables: the size of the scatterer, its refractive index, and the light wavelength.

2.4.2 Transport Mean Free Path, l_t

The transport mean free path, l_t is defined as the average distance that the light travels before its direction of propagation is randomized. It is defined as

$$l_t = \frac{l_s}{1 - g}, \tag{2.7}$$

where $g = \langle \cos\theta \rangle$ is the anisotropy parameter, θ is the scattering angle for a single scattering event in the sequence, and $\langle \ldots \rangle$ indicates an ensemble average over all the scattering events of the sequence. For isotropic scattering, $g = 0$.

2.4.3 Gain Length, l_g and Amplification Length, l_{amp}

Light amplification by stimulated emission in random media is described by the gain length l_g and the amplification length l_{amp}. The gain length is defined as the path length in the active medium over which light intensity is amplified by a factor *e*. The amplification length is defined as the average distance between the beginning and ending point for paths of length l_g. When the medium is homogeneous without scattering $l_{\text{amp}} = l_g$. Thus,

$$l_{\text{amp}} = \sqrt{\frac{l_t l_g}{3}}. \tag{2.8}$$

2.4.4 Distinction Between Feedback Regime by the Characteristic Length Parameters

Lasers with non-resonant feedback is a terminology first proposed in the late 1960s. The feedback in such lasers is due to return of part of the photons energy to the gain medium. The light in the cavity suffers multiple scattering, changing its propagation direction each time it is scattered; thus, any spatial resonance is absent. The stability condition due to the transport mean free path is $L \gg l_t \gg \lambda$, and this condition defines a ***diffusive regime***. In this type of RL, the phase of the optical field within the scattering medium is ignored because of the non-resonant nature of feedback provided by the scatterers. Consequently, the mean frequency of the RL only depends on the center frequency of the emission band of the gain media. Therefore, this kind of laser is called a RL with non-resonant feedback.

On the other hand, the coherent feedback is characterized by strong scattering, which is achieved when the transport mean free path becomes comparable with the emission wavelength (***localization regime***, $l_t \sim \lambda$). If the amplification along such a loop path exceeds the losses, the oscillation of the laser should occur in the loop which serves as a "laser resonator." In this case, the phase shift along the loop is a multiple of 2π (condition for constructive interference). Random lasing with coherent feedback is characterized by the appearance of narrow discrete laser peaks in the emission spectrum (spikes) in addition to a drastic increase in the emission intensity.

2.5 Modes in Random Lasers

Even in the absence of a conventional optical cavity, there is experimental evidence to support resonant optical modes unveiled by the presence of narrow peaks in the RL spectra, as described in Section 2.3. In order to explain the spikes from a quantitative point of view, Cao et al. [25, 30] calculated the electromagnetic field distribution in the random medium by solving Maxwell's equations considering the optical gain through a negative conductance, $\sigma(\omega)$, and the spontaneous emission as a noise current. The Maxwell's equations were solved in the time domain, and the output spectra were obtained through the Fourier transform with the output intensity given by $I(\omega) \propto |E(\omega)|^2$. The authors showed, by increasing the optical gain, the emergence of a short peak in the frequency domain, and the presence of bright spots (spikes) spatially distributed near the center of the random medium (Fig. 2.5), attributed to resonant longitudinal modes in the disordered medium associated with the closed loop paths for light.

Jiang and Soukoulis [31] presented a semi-classical theory based on the coupling between Maxwell's equations and the rate equations for the populations of the energy levels of the emitting system

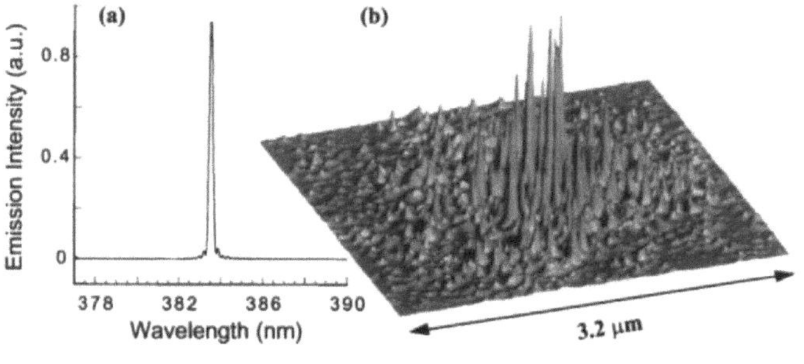

Figure 2.5 Simulated spectrum (a) and spatial distribution of emission intensity (b) considering the phase of the optical fields in random lasers. From Ref. [25], with permission.

(Maxwell–Bloch's equations).

$$\nabla \times \boldsymbol{E} = -\frac{\partial \boldsymbol{B}}{\partial t}$$
$$\nabla \times \boldsymbol{H} = \varepsilon \frac{\partial \boldsymbol{E}}{\partial t} + \frac{\partial \boldsymbol{P}}{\partial t}, \tag{2.9}$$

where $\boldsymbol{B} = \mu \boldsymbol{H}$, and the gain is introduced by the polarization vector, $\boldsymbol{P}$. Considering a single electron in the presence of an electric field, the polarization $P(x, t)$ is given by

$$\frac{d^2 P(t)}{dt^2} + \Delta\omega_a \frac{dP(t)}{dt} + \omega_a^2 P(t) = \frac{\gamma_r}{\gamma_c} \frac{e^2}{m} \Delta N(t) E(t), \tag{2.10}$$

where $\Delta\omega_a = \tau_{21}^{-1} + 2T_2^{-1}$ is the full width at half maximum of the atomic transition, $T_2$is the average time between dephasing events, $\gamma_r = \tau_{21}^{-1}$ is the real decay rate of level 2, and γ_c is the classical rate. The gain medium is modeled as a four-level system with populations N_0, N_1, N_2, and N_3, see Eq. 2.11. The electrons are excited from the ground state, 0, to the level 3 by an excitation source operating at a rate P_r. Once excited, the electrons at level 3 can relax to the lower level 2 at a rate τ_{32}^{-1}. The populations of the upper and lower levels of the RL transition are N_2 and N_1, respectively. The relaxation from level 2 to level 1 is due to spontaneous (rate τ_{21}^{-1}) and stimulated emissions. The relaxation from the level 1 to the level 0 is nonradiative at a rate τ_{10}^{-1}. The RL frequency is given by $\omega_a = (E_2 - E_1)/\hbar$, where E_2 and E_1 are the energies associated with levels 2 and 1, respectively. In the equation above, $\Delta N(x, t) = N_1(x, t) - N_2(x, t)$. The populations of the levels are determined by solving the set of Eq. 2.11.

$$\begin{aligned}
\frac{dN_3(x,t)}{dt} &= P_r N_0(x,t) - N_3(x,t)\tau_{32}^{-1} \\
\frac{dN_2(x,t)}{dt} &= N_3(x,t)\tau_{32}^{-1} + \frac{E(x,t)}{\hbar\omega_a}\frac{dP(x,t)}{dt} - N_2(x,t)\tau_{21}^{-1} \\
\frac{dN_1(x,t)}{dt} &= N_2(x,t)\tau_{21}^{-1} - \frac{E(x,t)}{\hbar\omega_a}\frac{dP(x,t)}{dt} - N_1(x,t)\tau_{10}^{-1} \\
\frac{dN_0(x,t)}{dt} &= N_1(x,t)\tau_{10}^{-1} - P_r N_0(x,t).
\end{aligned} \tag{2.11}$$

The stimulated emission or absorption at the RL frequency is determined by the sign of $\frac{E(x,t)}{\hbar\omega_a}\frac{dP(x,t)}{dt}$. Considering a 1D RL geometry, the authors demonstrated the existence of an excitation power

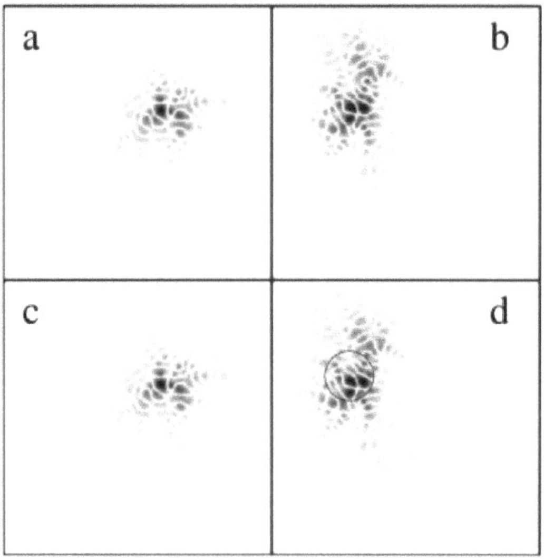

Figure 2.6 Spatial distribution of the magnitude of the laser field under selective excitation of two given modes in a random medium without gain (passive system): (a) mode $\lambda = 446.9$ nm, and (b) mode $\lambda = 445.0$ nm. (c) and (d) represent the distribution of the magnitude of the laser field under extended and spatially localized gain, respectively. In (d) the circle represents the spatial extension of the gain. From Ref. [26], with permission.

threshold for RL emission, as well as an increasing of the number of modes with the excitation power and the size of the system. The results are consistent with the experimental observations [21, 22, 25, 30]. Similar approach was adopted by Vanneste and Sebbah [26, 28], who demonstrated in two dimensions that the optical modes are determined by the scatterers configuration, the presence of gain does not alter the optical modes, and the output spectrum depends on the location of the excitation source (Fig. 2.6). This last result indicates that the output spectrum, including the number of modes, could be controlled by selective excitation of the random medium.

Remarkably, Vanneste et al. [32] demonstrated that, even in weakly scattering RLs, far from the Anderson localization regime, it is still possible to get resonant feedback, in which interference effects are unveiled by the presence of sharp peaks in the RL emission spectra: one peak for gain just above the threshold and

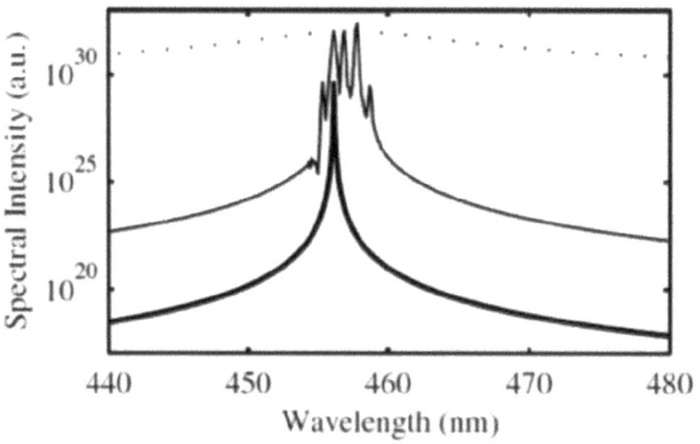

Figure 2.7 Emission spectrum in the steady state regime for a random laser in the weakly scattering regime with one mode (thick line) and with several modes at higher pumping rate (thin line). The gain profile is represented by the dotted line. Adapted from Ref. [32], with permission.

several peaks at higher gain (Fig. 2.7). The demonstration was based on a regime between ballistic and diffusive, far from the localization regime, by coupling Maxwell's equations with the rate equations of a four-level atomic system. It was found that the lasing modes consist of traveling waves spatially extended over the whole system. The authors presented evidence that the modes correspond to the quasimodes of the passive system, and in contrast with localized regimes in which the modes are standing waves, they have traveling components, meaning that the fields are complex.

In order to describe the full nonlinear dynamics of RLs, Andreasen et al. [33] simulated the temporal evolution of 1D and 2D RLs using an approach similar to that of Jiang and Soukoulis [31] described above, in order to get the solution of the Maxwell–Bloch equations using the finite-difference time-domain method. Andreasen et al. [33] unveiled several manifestations of optical nonlinearities, such as strong relaxation oscillations, mode competition, mode suppression, and, by considering third-order nonlinear effects, demonstrated the first theoretical predictions of third-harmonic generation, four-wave mixing, and sum-frequency generation. Indeed, effects arising from second-order optical nonlinearity have been demonstrated experimentally in Nd^{3+} based RLs, as second-harmonic [34–37], sum-frequency [35–37], and difference-frequency [38] generations.

More recently, Türeci et al. [39–41] presented a theoretical formulation for diffusive RLs based on semiclassical laser theory considering biorthogonal modes, called constant flux (CF) states. This approach is based on the Maxwell–Bloch equations, which are coupled nonlinear equations for the electric field, polarization, and population inversion of the gain medium. In the stationary state, the modes are represented by

$$\Psi_\mu(\mathbf{x}) = \frac{i\gamma_\perp}{\gamma_\perp - i(k_\mu - k_a)} \int \frac{D_0(\mathbf{x})G(\mathbf{x}, \mathbf{x}'; k_\mu)\Psi_\mu(\mathbf{x}')}{\varepsilon(\mathbf{x}')(1 + \Sigma_\nu \Gamma_\nu |\Psi_\mu(\mathbf{x}')|^2)} d\mathbf{x}', \quad (2.12)$$

and the electric field reads $e(\mathbf{x}, t) = \Sigma_\nu \Psi_\mu(\mathbf{x}) e^{-i\Omega_\mu t}$. The lasing frequencies are $\Omega_\mu = ck_\mu$, k_a is the atomic frequency, $\gamma_\perp$ is the transverse relaxation rate, Γ_ν is the gain profile, $D_0(\mathbf{x})$ is the spatial distribution of the pump intensity, and $\varepsilon(\mathbf{x}')$ is the dielectric function of the "resonator." Nonlinear interaction among the optical modes is related to the denominator of Eq. 2.12, as well as mode competition, in the sense that one mode can suppress the others through spatial hole-burning. The integral is over the gain region and the kernel $G(\mathbf{x}, \mathbf{x}'; k_\mu)$ is the Green function of the cavity with purely outgoing boundary conditions, that represents the steady-state response of the passive cavity to a monochromatic electric dipole oscillating at frequency k at $\mathbf{x}'$. $G(\mathbf{x}, \mathbf{x}'; k_\mu)$ is represented in terms of dual sets of biorthogonal functions $\varphi_m(\mathbf{x}, k)$ and $\bar{\varphi}_m(\mathbf{x}, k)$, which are the constant flux states, with complex eigenvalues k_m. In this approach, the constant flux states are the resonances of the linear cavity and reduce to the quasi-bound states for high-Q resonators. By considering a given disorder configuration, the intensities of the optical modes are shown in Fig. 2.8 as a function of the pumping intensity. Mode interaction is observed when two given modes approach in frequency (inset of Fig. 2.8). For the modes represented by orange and black, one can see that the first increases at the cost that the other drops to zero at large pump strength. In the inset, the frequencies above threshold are given by solid lines, while the ones below the threshold are represented by dashed lines.

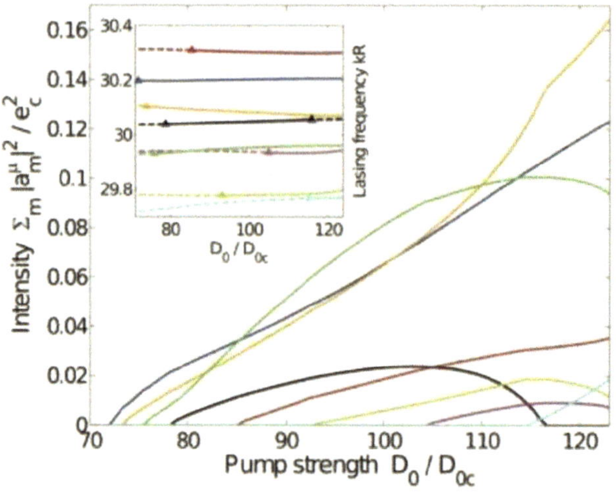

Figure 2.8 Modal intensities of a diffusive random laser versus pump strength D_0 for a given disorder configuration. The inset shows the frequencies of the lasing modes both above (solid) and below (dashed) threshold versus D_0. The frequencies can cross if one mode is not lasing but not if both are lasing, i.e., modes with nearby frequencies interact strongly and their intensity curves are anticorrelated. From Ref. [41], with permission.

The strong interaction among nearby frequencies is unveiled by the anti-correlation in the intensity curves, i.e., when one is lasing the other is not. The spaced lasing frequencies in the diffusive RLs are attributed to the spatial overlap of the quasi-degenerate modes and the mode competition, with strong hole-burning, which suppresses some of the lasing frequencies. The situation is quite different for larger disorder strength, when the mode interaction decreases as one approaches the localization regime.

In the approach of Türeci et al., instead of solving the time-dependent Maxwell–Bloch equations, they introduced an ab initio self-consistent formalism, which is a time-independent method to find the stationary properties of the system [39–43]. This is a powerful tool able to predict the outcomes by providing only few inputs such as the dielectric function of the host medium and two fundamental parameters of the gain medium (frequency of the lasing transition and its relaxation rate). However, the cost to pay in this approach is that the full dynamic is lost.

2.6 Are the Spikes Evidence of Resonant Feedback?

In spite of the distinction of the feedback regime by the observation or not of narrow peaks in the RL spectra, recent experimental results demonstrated that the so called non-resonant feedback may be due to the superposition of a large number of optical modes resulting in smooth output spectra, i.e., spikeless due to a self-average mechanism (Fig. 2.9) [44, 45]. The narrow peaks could be observed by using high resolution spectral and time measurements, and each mode is stochastically activated from one experimental realization to another under the same initial conditions Spikes suppression due to self-averaging are more easily observed by using long excitation beam pulses, and large excitation volumes [44, 45]. Equivalently, an averaging mechanism can be obtained by spectral accumulation over several shots of the excitation beam [46].

Although the spikes observed in some RLs are supported by theoretical models as manifestations of interference effects, they can also be described by the diffusion model, when using Monte Carlo simulations, as described in Section 2.7. However, a definite evidence of oscillation in RLs was given by the photon statistics: Coherent sources present Poisson distributions, and incoherent light Bose–Einstein distributions. A general overview on photon statistics can be found in [47]. In the first report on photon statistics of RLs, Zacharakis et al. [48] found that the photon statistics of a Rhodamine 6G based RL is constituted of a superposition of Poisson and Bose–Einstein distributions. Afterwards, Cao et al. [49] investigated ZnO nanoparticles and demonstrated Poisson distributions at large excitation power. Regarding the longitudinal and transversal coherences, Noginov et al. [50] found that the longitudinal coherence length of a Nd^{3+} based powder RL is as large as 1.7 cm. However, no transversal coherence length was detected. Papadakis et al. [51] investigated Rhodamine 6G based RL, and

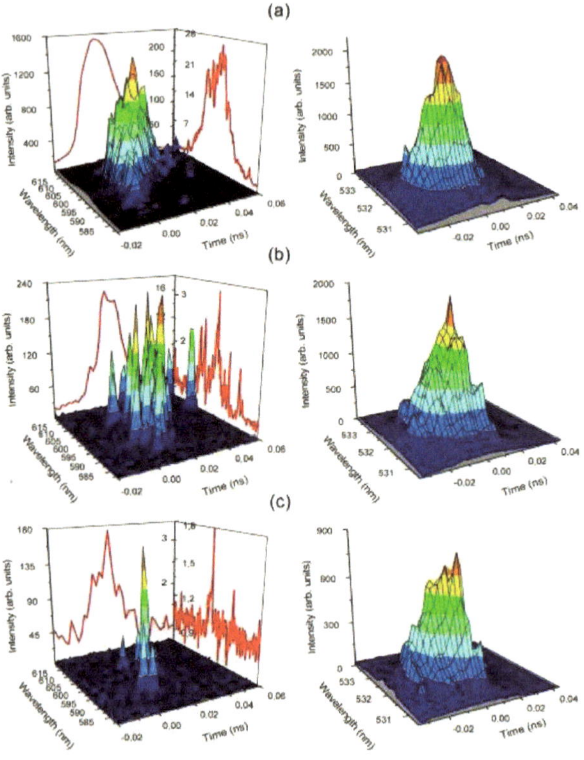

Figure 2.9 3D map surfaces of the random laser (left) and excitation pulses (right) obtained with pin holes as spatial filters with diameters 50 μm (a), 30 μm (b), and 15 μm (c) while keeping constant the experimental conditions, but in different experimental realizations (shots). From Ref. [44], with permission.

demonstrated that the longitudinal coherence length increased up to ~17 μm by increasing the excitation power.

2.7 Description of Spikes without Casting the Phase of the Optical Fields

By using Monte Carlo simulations [52, 53], the spikes can be described by amplification of photons in random walk through the

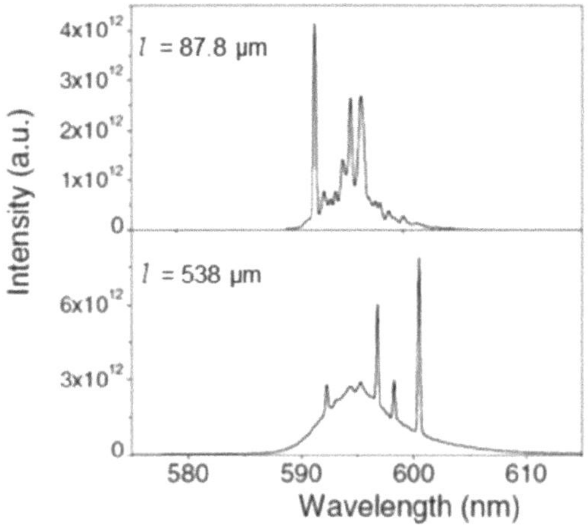

Figure 2.10 Monte Carlo simulation of the emission spectra for different transport mean free path, *l*. Sharp emission peaks are reproduced, which are associated with rare long light paths that were amplified exponentially. From Ref. (54), with permission.

disordered medium neglecting the phases of the fields [54, 55]. As first demonstrated by Mujumdar et al. [54], the introduction of exponential gain in a multiple light scattering process strongly increases the importance of rare very long light paths dominating the emission spectrum with large peak intensities (Fig. 2.10). This was also corroborated by an analytical model [56], which describes the emergence of rare large intensity spikes considering that the intensity of a given wavelength increases with the path length, l, as $I(l) = I_0 \exp(l/l_g)$, and the probability of a given length decreases with l as $P(l) = \exp(-l/\langle l\rangle)/\langle l\rangle$, with $\langle l\rangle$ being the mean free path of photons. As demonstrated by Ignest et al. [55], when the scattering strength is low, the "lucky" photons can take the gain and the corresponding wavelength appears as a large peak in the output spectrum. On the other hand, increasing the scattering strength, the spikes disappear, and one observes only a spectral narrowing when increasing the optical gain. The spikes suppression is attributed to the spatial superposition and temporal coincidence of a large number of optical modes which depletes the available gain leading to self-averaging spectral distributions. The experimental results were well described by Monte Carlo simulations, for the photon dynamics in the disordered medium. In this approach, the particle wave of light was considered, and then interference effects were not unveiled.

2.8 Speckle in Random Lasers

One remarkable characteristic of laser light is the capability to produce speckle. It is a direct manifestation of the wave nature of light, and can be easily observed thanks to the large coherence length that some laser sources can present. From a technological point of view, more specifically for image systems, the speckle has to be suppressed to generate high quality images. However, it was demonstrated that RLs are able to produce speckle-free images due to the spatial and temporal superposition of a large number of optical modes [57]. In Ref. [58], the authors measured the number of modes present in a RL emission by the speckle contrast (C), taking the advantage that it is a measure of the coherence of the light source. In other words, speckle contrast is a measure of how strong the intensity fluctuations in a speckle pattern are, compared to the average intensity, and can be evaluated by the ratio of the standard deviation (σ_I) over the speckle average intensity ($\langle I \rangle$). Under the assumption that the average intensities are equal, i.e., $\langle I \rangle = I_0$ for all n, this result reduces to the expression $C = N^{-1/2}$, where N is the number of independent speckle patterns [59]. It may also correspond to the independent number of RL modes.

In Ref. [60], the speckle contrast was used to measure the number of modes present in the laser emission of a random laser and a random fiber laser (RFL), which is the one-dimensional fiber waveguide version of RLs. The narrowest Er-RFL spectrum measured was limited by the instrument resolution of 0.06 nm (Fig. 2.11), and therefore does not show spikes representative of the longitudinal modes. However, the multimodal nature of the Er-RFL was identified by the speckle pattern with determination of more than 200 modes. Figure 2.12 shows the results for the number of modes for all lasers studied in [60], including the excitation lasers.

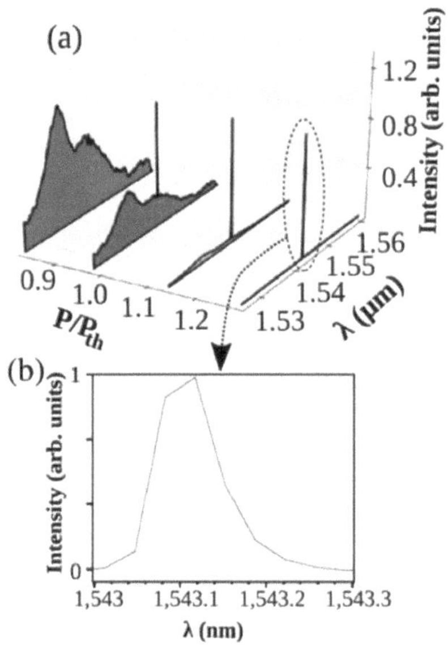

Figure 2.11 (a) Spectral evolution as function of the pump power P, normalized by threshold power P_{th}. (b) Resolution limited of lasing spectrum, and therefore does not show spikes representative of the longitudinal modes. From Ref. [60], with permission.

2.9 Output Intensity Fluctuation of Random Lasers

One intriguing characteristic observed in RLs is the output intensity fluctuations. Several manuscripts have been showing that in experimental realizations under the same initial conditions the output spectra from one realization to another can differ significantly. This characteristic was first analyzed by the probability distributions for the output intensity fluctuations, and it was showed that for gain smaller than the RL threshold gain, the distributions are Gaussian, and Lévy for gain above the threshold [61]. Other articles investigated it in details. Remarkably, Ignesti et al. [55] performed Monte Carlo simulations considering as random parameters the spontaneous emission, and the diffusion of the photons. The randomness of the spontaneous emission was due to the initial

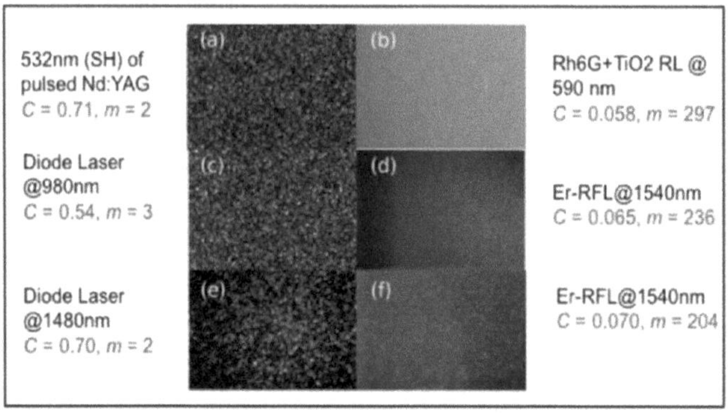

Figure 2.12 Speckle images of (a) second harmonic (532 nm) of a pulsed Nd:YAG laser, (b) Rh6G-TiO_2 RL (590 nm), (c) 980 nm semiconductor laser, (d) Er-RFL pumped at 980 nm, (e) 1480 nm semiconductor laser, and (f) Er-RFL pumped at 1480 nm. Adapted from Ref. [60], with permission.

direction of each spontaneously created photon, and emitted wavelength which was weighted by the spontaneous emission around the resonance transition. The simulation reproduced the spikes observed in some RLs, and the crossovers between statistical regimes. The essence of both effects was the competition between the modes for the available gain. The spikes were attributed to the "lucky" photons, which take the gain suppressing photons at other wavelengths to be amplified. It was also the reason for the first crossover from Gaussian to Lévy distributions at the onset of RL emission. Increasing the gain beyond the threshold gain, there is a return to the Gaussian regime which was attributed to the spatial superposition and temporal coincidence of a large number of competing optical modes which suppress large intensities in the output spectra. One can think it as a self-averaging system, and the spikes emerge when the scattering strength is low. In these cases, there is no Gaussian–Lévy crossover meaning that the probability distributions for the output fluctuations are Gaussian, independent of the optical gain amplitude. We recall that even analyzing the stationary state, Türeci et al. [41] predicted mode competition when considering the wave nature of light, as described in Section 2.5.

Other approach is based on the analogy between the RL modes and the spins of magnetic systems. Angelani et al. [62,

63] investigated the nonlinear interaction of the optical modes via third order nonlinearity of the host medium, and made an analogy between the optical modes and the spins of magnetic systems. The authors predicted a replica symmetry breaking (RSB) by increasing the optical gain (for more details, see Chapter 5). The RSB is a concept of the spin glass theory, which states that systems prepared under identical initial conditions can yield different results of observable quantities. Although this is a concept from the spin glass, it was only experimentally demonstrated in 2015 using a RL system [64], as described in detail in Chapter 6. Remarkably, in Chapter 5 the simultaneous observation of both the Lévy distributions and the replica symmetry breaking in a single set of measurements is predicted and demonstrated experimentally in Chapter 6.

2.10 Conclusion

Although the random lasers (RLs) do not have an optical cavity to support resonant modes, the emitted light is quite complex. In the earliest investigations, two types of feedback were attributed by the characteristic of the output spectra displaying short narrow peaks (resonant or coherent feedback) or not (non-resonant or incoherent feedback). However, it is a consensus nowadays that regardless of the direct manifestation of narrow peaks in the output spectra or not, the RLs possess optical modes which can be accessed by high resolution spectro-temporal measurements, and by image contrast. One puzzling characteristic of the output spectra is the output intensity fluctuations, which have been investigated within the framework of the statistical mechanics. The optical modes are stochastically activated from one experimental realization to another under the same initial conditions, and competes for the available gain emerging quite interesting experimental observations, as Gaussian or Lévy distributions for the intensity fluctuations. Also, the optical modes interact via nonlinear response of the optical medium, and analogies with spin-glasses were proposed and experimentally demonstrated, as discussed in the following Chapters.

Acknowledgments

This work was supported by the Brazilian agencies Conselho Nacional de Desenvolvimento Cientfico e Tecnológico (CNPq), Financiadora de Estudos e Projetos (Finep), and Fundação de Amparo à Ciência e Tecnologia do Estado de Pernambuco (FACEPE).

References

1. Moura, A. L., Fewo, S. I., Carvalho, M. T., Kuzmin, A. N., Prasad, P. N., Gomes, A. S. L., and de Araújo, C. B. (2015). Random lasing in Nd^{3+} doped potassium gadolinium tungstate crystal powder, *J. Appl. Phys.*, **117,** 083102.
2. Lawandy, N. M., Balachandran, R. M., Gomes, A. S. L., and Sauvain, E. (1994). Laser action in strongly scattering media, *Nature,* **368,** 436–438.
3. de Matos, C. J. S., Menezes, L. D. S., Brito-Silva, A. M., Gamez, M. A. M., Gomes, A. S. L., and de Araújo, C. B. (2007). Random fiber laser, *Phys. Rev. Lett.,* **99**, 153903.
4. Song, Q. H., Xiao, S. M., Xu, Z. B., Liu, J. J., Sun, X. H., Drachev, V., Shalaev, V. M., Akkus, O., and Kim, Y. L. (2010). Random lasing in bone tissue, *Opt. Lett.,* **35**, 1425–1427.
5. Polson, R. C., and Vardeny, Z. V. (2010). Cancerous tissue mapping from random lasing emission spectra, *J. Opt.,* **12**, 024010.
6. Letokhov, V. S. (1967). Stimulated emission of an ensemble of scattering particles with negative absorption, *JETP Lett.,* **5**, 212–215.
7. Letokhov, V. S. (1968). Generation of light by a scattering medium with negative resonance absorption, *Sov. Phys. JETP,* **26**, 835–840.
8. Davison, B., and Sykes, J. B. (1958). *Neutron Transport Theory,* Oxford University Press, Oxford.
9. Fork, R. L., Taylor, D. W., German, K. R., Kiel, A., and Buehler, E. (1974). Unusual luminescence from crystals containing Eu^{2+}, *Phys. Rev. Lett.,* **32**, 781–783.
10. Varsanyi, F. (1971). Surface lasers, *Appl. Phys. Lett.,* **19**, 169–171.
11. Markushev, V. M., Zolindaki, A., and Briskina, C. M. (1986). Luminescence and stimulated emission of neodymium in sodium lanthanum molybdate powders, *Sov. J. Quantum Electron.,* **16**, 281–283.

12. Qian, S.-X., Snow, J. B., Tzeng, H.-M., and Chang, R. K. (1986). Lasing droplets: highlighting the liquid-air interface by laser emission, *Science,* **231**, 486–488.
13. Wiersma, D. S., Vanalbada, M. P., and Lagendijk, A. (1995). Random laser?, *Nature,* **373**, 203–204.
14. Pincheira, P. I. R., Silva, A. F., Fewo, S. I., Carreño, S. J. M., Moura, A. L., Raposo, E. P., Gomes, A. S. L., and de Araújo, C. B. (2016). Observation of photonic paramagnetic to spin-glass transition in a specially designed TiO_2 particle-based dye-colloidal random laser, *Opt. Lett.,* **41**, 3459–3462.
15. Wiersma, D. S., and Lagendijk, A. (1996). Light diffusion with gain and random lasers, *Phys. Rev. E,* **54**, 4256.
16. Noginov, M. A., Noginova, N. E., Caulfield, H. J., Venkateswarlu, P., Thompson T., Mahdi, M., and Ostroumov, V. (1996). Short-pulsed stimulated emission in the powders of $NdAl_3(BO_3)_4$, $NdSc_3(BO_3)_4$, and $Nd{:}Sr_5(PO_4)_3F$ laser crystals, *J. Opt. Soc. Am. B.,* **13**, 2024–2033.
17. Moura, A. L., Maia, L. J., Jerez, V., Gomes, A. S., and de Araújo, C. B. (2019). Random laser in $Nd{:}YBO_3$ nanocrystalline powders presenting luminescence concentration quenching, *J. Lumin.,* **214**, 116543.
18. Moura, A. L., Pincheira, P. I. R., Reyna, A. S., Raposo, E. P., Gomes, A. S. L., and de Araújo, C. B. (2017). Replica symmetry breaking in the photonic ferromagneticlike spontaneous mode-locking phase of a multimode Nd: YAG laser, *Phys. Rev. Lett.,* **119**, 163902.
19. Pincheira, P. I., da Silva-Neto, M. L., Maldonado, M., de Araújo, C. B., Jawaid, A. M., Busch, R., Ritter, A. J., Vaia, R. A., and Gomes, A. S. (2020). Monolayer 2D $ZrTe_2$ transition metal dichalcogenide as nanoscatter for random laser action, *Nanoscale,* **12**, 15706–15710.
20. Cao, H., Zhao, Y. G., Ong, H. C., Ho, S.-T., Dai, J. Y., Wu, J. Y., and Chang, R. P. (1998). Ultraviolet lasing in resonators formed by scattering in semiconductor polycrystalline films, *Appl. Phys. Lett.,* **73**, 3656–3658.
21. Cao, H., Zhao, Y. G., Ho, S. T., Seelig, E. W., Wang, Q. H., and Chang, R. P. H. (1999). Random laser action in semiconductor powder, *Phys. Rev. Lett.,* **82**, 2278–2281.
22. Frolov, S. V., Vardeny, Z. V., Yoshino, K., Zakhidov, A., and Baughman, R. H. (1999). Stimulated emission in high-gain organic media, *Phys. Rev. B,* **59**, R5284.
23. Cao, H., Xu, J. Y. Y., Ling, Y., Burin, A. L., Seeling, E. W., Liu, X., and Chang, R. P. H. (2003). Random lasers with coherent feedback, *IEEE J. Sel. Top. Quant. Electron.,* **9**, 111–119.

24. Anderson, P. W. (1958). Absence of diffusion in certain random lattices, *Phys. Rev.*, **109**, 1492–1505.
25. Cao, H., Xu, J. Y., Zhang, D. Z., Chang, S.-H., Ho, S. T., Seelig, E. W., Liu, X. and Chang, R. P. (2000). Spatial confinement of laser light in active random media, *Phys. Rev. Lett.*, **84**, 5584.
26. Vanneste, C., and Sebbah, P. (2001). Selective excitation of localized modes in active random media, *Phys. Rev. Lett.*, **87**, 183903.
27. Sánchez-Gil, J. A., and Freilikher, V. (2003). Local and average fields inside surface-disordered waveguides: resonances in the one-dimensional Anderson localization regime, *Phys. Rev. B*, **68**, 075103.
28. Sebbah, P., and Vanneste, C. (2002). Random laser in the localized regime, *Phys. Rev. B*, **66**, 144202.
29. Jiang, X., and Soukoulis, C. M. (2002). Localized random lasing modes and a path for observing localization, *Phys. Rev. E*, **65**, 025601.
30. Cao, H., Xu, J. Y., Chang, S. H., and Ho, S. T. (2000). Transition from amplified spontaneous emission to laser action in strongly scattering media, *Phys. Rev. E*, **61**, 1985–1989.
31. Jiang, X., and Soukoulis, C. M. (2000). Time dependent theory for random lasers, *Phys. Rev. Lett.*, **85**, 70–73.
32. Vanneste, C., Sebbah, P., and Cao, H. (2007). Lasing with resonant feedback in weakly scattering random systems, *Phys. Rev. Lett.*, **98**, 143902.
33. Andreasen, J., Sebbah, P., and Vanneste, C. (2011). Nonlinear effects in random lasers, *J. Opt. Soc. Am. B*, **28**, 2947–2955.
34. Noginov, M. A., Egarievwe, S. U., Noginova, N., Wang, J. C., and Caulfield, H. J. (1998). Demonstration of a second-harmonic powder laser, *J. Opt. Soc. Am. B*, **15**, 2854–2860.
35. Moura, A. L., Jerez, V., Maia, L. J. Q., Gomes, A. S. L., and de Araújo, C. B. (2015). Multi-wavelength emission through self-induced second-order wave-mixing processes from a Nd^{3+} doped crystalline powder random laser, *Sci. Rep.*, **5**, 13816.
36. Carreño, S. J. M., Moura, A. L., Fabris, Z. V., Maia, L. J. Q., Gomes, A. S. L., and de Araújo, C. B. (2016). Interplay between random laser performance and self-frequency conversions in $Nd_xY_{1.00-x}Al_3(BO_3)_4$ nanocrystals powders, *Opt. Mat.*, **54**, 262–268.
37. Moura, A. L., Carreño, S. J. M., Pincheira, P. I. R., Fabris, Z. V., Maia, L. J. Q., Gomes, A. S. L., and de Araújo, C. B. (2016). Tunable ultraviolet and blue light generation from Nd:YAB random laser bolstered by second-order nonlinear processes, *Sci. Rep.*, **6**, 27107.

38. Rocha, E. G., Paz, I. R., Santos, B. J., Soares, W. C., de Lima, E., Leão, L. M., Maia, L. J., and Moura, A. L. (2019). Self-induced optical parametric amplification of random laser emission, *Laser Phys.*, **29**, 045402.
39. Türeci, H. E., Stone, A. D., and Collier, B. (2006). Self-consistent multimode lasing theory for complex or random lasing media, *Phys. Rev. A,* **74**, 043822.
40. Türeci, H. E., Stone, A. D., and Ge, L. (2007). Theory of the spatial structure of nonlinear lasing modes, *Phys. Rev. A,* **76**, 013813.
41. Tureci, H. E., Ge, L., Rotter, S., and Stone, A. D. (2008). Strong interactions in multimode random lasers, *Science,* **320**, 643–646.
42. Ge, L., Tandy, R. J., Stone, A. D., and Türeci, H. E. (2008). Quantitative verification of ab initio self-consistent laser theory, *Opt. Express,* **16**, 16895–16902.
43. Türeci, H. E., Stone, A. D., Ge, L., Rotter, S., and Tandy, R. J. (2008). Ab initio self-consistent laser theory and random lasers, *Nonlinearity,* **22**, C1.
44. Garcia-Revilla, S., Fernandez, J., Barredo-Zuriarrain, M., Carlos, L. D., Pecoraro, E., Iparraguirre, I., Azkargorta, J., and Balda, R. (2015). Diffusive random laser modes under a spatiotemporal scope, *Opt. Express,* **23**, 1456–1469.
45. García-Revilla, S., Fernández, J., Barredo-Zuriarrain, M., Carlos, L. D., Pecoraro, E., Iparraguire, I., Azkargorta, J., and Balda, R. (2015). Spectral dynamics of a diffusive random laser under two photon pumping, *Adv. Device Mater.,* **1**, 38–45.
46. Gomes, A. S. L., Carvalho, M. T., Dominguez, C. T., de Araújo, C. B., and Prasad, P. N. (2014). Direct three-photon excitation of upconversion random laser emission in a weakly scattering organic colloidal system, *Opt. Express,* **22**, 14305–14310.
47. Klafter, J., Shlesinger, M. F., and Zumofen, G. (1996). Beyond Brownian motion, *Physics Today,* **49**, 33–39.
48. Zacharakis, G., Papadogiannis, N. A., Filippidis, G., and Papazoglou, T. G. (2000). Photon statistics of laserlike emission from polymeric scattering gain media, *Opt. Lett.,* **25**, 923–925.
49. Cao, H., Ling, Y., Xu, J. Y., Cao, C. Q., and Kumar, P. (2001). Photon statistics of random lasers with resonant feedback, *Phys. Rev. Lett.,* **86**, 4524–4527.
50. Noginov, M. A., Egarievwe, S. U., Noginova, N., Caulfield, H. J., and Wang, J. C. (1999). Interferometric studies of coherence in a powder laser, *Opt. Mat.,* **12**, 127–134.

51. Papadakis, V. M., Stassinopoulos, A., Anglos, D., Anastasiadis, S. H., Giannelis, E. P., and Papazoglou, D. G. (2007). Single-shot temporal coherence measurements of random lasing media, *J. Opt. Soc. Am. B,* **24**, 31–36.
52. Berger, G. A., Kempe, M., and Genack, A. Z. (1997). Dynamics of stimulated emission from random media, *Phys. Rev. E,* **56**, 6118–6122.
53. Siddique, M., Alfano, R. R., Berger, G. A., Kempe, M., and Genack, A. Z. (1996). Time-resolved studies of stimulated emission from colloidal dye solutions, *Opt. Lett.,* **21**, 450–452.
54. Mujumdar, S., Ricci, M., Torre, R., and Wiersma, D. S. (2004). Amplified extended modes in random lasers, *Phys. Rev. Lett.,* **93**, 053903.
55. Ignesti, E., Tommasi, F., Fini, L., Lepri, S., Radhalakshmi, V., Wiersma, D., and Cavalieri, S. (2013). Experimental and theoretical investigation of statistical regimes in random laser emission, *Phys. Rev. A,* **88**, 033820.
56. Lepri, S., Cavalieri, S., Oppo, G. L., and Wiersma, D. S. (2007). Statistical regimes of random laser fluctuations, *Phys. Rev. A,* **75**, 063820.
57. Redding, B., Choma, M. A., and Cao, H. (2012). Speckle-free laser imaging using random laser illumination, *Nat. Photonics,* **6**, 355–359.
58. Hokr, B. H., Cerjan, A., Thompson, J. V., Yuan, L., Liew, S. F., Bixler, J. N., Noojin, G. D., Thomas, R. J., Cao, H., Stone, A. D., Rockwell, B. A., Scully, M. O., and Yakovlev, V. V. (2016). Evidence of Anderson localization effects in random Raman lasing, in *SPIE LASE* (Vodopyanov, K. L., and Schepler, K. L., eds), San Francisco, California, United States, pp. 973110.
59. Goodman, J. W. (2020). *Speckle Phenomena in Optics: Theory and Applications,* 2nd ed, SPIE Press Book.
60. Gomes, A. S. L., Lima, B. C., Pincheira, P. I. R., Moura, A. L., Gagné, M., Raposo, E. P., de Araújo, C. B., and Kashyap, R. (2016). Glassy behavior in a one-dimensional continuous-wave erbium-doped random fiber laser, *Phys. Rev. A,* **94**, 011801(R).
61. Sharma, D., Ramachandran, H., and Kumar, N. (2006). Lévy statistical fluctuations from a random amplifying medium, *Fluct. Noise Lett.,* **6**, L95–L101.
62. Angelani, L., Conti, C., Ruocco, G., and Zamponi, F. (2006). Glassy behavior of light, *Phys. Rev. Lett.,* **96**, 065702.
63. Angelani, L., Conti, C., Ruocco, G., and Zamponi, F. (2006). Glassy behavior of light in random lasers, *Phys. Rev. B,* **74**, 104207.
64. Ghofraniha, N., Viola, I., Di Maria, F., Barbarella, G., Gigli, G., Leuzzi, L., and Conti, C. (2015). Experimental evidence of replica symmetry breaking in random lasers, *Nat. Commun.,* **6**, 6058.

Chapter 3

Analytical Approaches to Lévy Statistics in Random Lasers: A Comparison with Numerical and Experimental Results, Extreme Events, and the Influence of High-Order Nonlinearities on the Intensity Fluctuations

Ernesto P. Raposo[a] and Anderson S. L. Gomes[b]

[a] *Laboratório de Física Teórica e Computacional, Departamento de Física, Universidade Federal de Pernambuco, 50670-901 Recife-PE, Brazil*

[b] *Departamento de Física, Universidade Federal de Pernambuco, 50670-901 Recife-PE, Brazil*

ernesto.raposo@ufpe.br

3.1 Introduction

The deepening of the investigations on the statistical properties of the output intensity fluctuations in random laser (RL) systems in the last decade revealed a rather striking property [1–19]: depending on

Lévy Statistics and Spin Glass Behavior in Random Lasers
Edited by Anderson S. L. Gomes, André L. Moura, Cid B. de Araújo, and Ernesto P. Raposo

ISBN 978-981-4968-55-3 (Hardcover), 978-1-003-33618-1 (eBook)
www.jennystanford.com

the excitation power regime, the distribution of the emitted intensity values presents Lévy-like properties.

The first experimental study on the RL intensity fluctuations was reported in Ref. [1], whereas the first theoretical insights appeared in Refs. [2, 4, 6]. The dependence of their statistical regimes on the input excitation power was first unveiled in [5, 6, 9, 11]. In particular, while in Ref. [6] (see also Refs. [13, 14]) Lepri and co-authors offered an explanation for the origin of Lévy-like behavior based on the statistical properties of the photon stochastic trajectories that are multiply scattered within the RL sample, in Ref. [11] Uppu and Mujumdar proposed the use of the exponent associated with the family of Lévy α-stable distributions as a universal identifier of the threshold and criticality in RLs by performing numerical simulations and experiments with dye-scatterer-based RLs. An overview of the seminal contribution by Uppu and Mujumdar on the Lévy-like properties of RLs can be found in Chapter 4 of this book. Later on, the same pattern of intensity fluctuations was experimentally demonstrated [15–19] in the RL emissions of an erbium-doped random fiber laser (RFL).

The aim of this chapter is to review the theoretical approaches that lead to the Lévy-like behavior of output intensity values in RL systems.

We start by briefly introducing in Section 3.2 the ideas by Lepri and co-authors that give rise to a power-law probability density function (PDF) of intensities, which represents the asymptotical behavior of the Lévy α-stable distribution. We next present in Section 3.3 a rather distinct approach for the Lévy-type PDF of intensities based on the Langevin and associated Fokker–Planck equations for the dynamics of the mode amplitudes with disordered nonlinear medium. The study of the contribution of higher-order nonlinearities to the PDF of intensities is presented in Section 3.4 and the analysis of extreme values of intensities is given in Section 3.5. Connections with the theory of the photonic replica-symmetry-breaking spin-glass phase of RL systems are discussed in Section 3.6. Finally, conclusion remarks are presented in Section 3.7.

3.2 Lévy Statistics in Random Lasers: The Stochastic Trajectory Approach

In the seminal work published in Ref. [6], Lepri and co-authors investigated analytically and numerically the origin of the experimentally observed power-law behavior of the distribution of output intensity values in RLs. They based their theory on the amplification of diffusing photons, which interact with the gain medium and get multiply scattered. We now briefly review their main result, i.e., the derivation of a power-law Lévy-type PDF of intensities in the stochastic trajectory approach.

Consider that a photon labeled i of wavelength λ exits the resonant cavity after traversing a first-passage random trajectory of length l. The amplification process within the cavity scales exponentially with the path length l, so that the associated emission intensity signal is given by [6]

$$I_i(l\lambda) = I_0 \exp(l/\ell_G), \tag{3.1}$$

where I_0 is its minimum value as the process starts with a single spontaneous emission photon and $\ell_G(\lambda)$ is the gain length of the active medium, i.e., the mean distance for the first amplification of the photon.

The disordered character of the medium leads the length l of the trajectory to be a random variable described by an exponential PDF,

$$p(l) = \frac{\exp(-l/\langle l \rangle)}{\langle l \rangle}, \tag{3.2}$$

with $\langle l \rangle(\lambda)$ denoting the average length of the photon paths with wavelength λ within the RL sample. By combining Eqs. (3.1) and (3.2), the cumulative distribution of the intensity of the exiting photons is therefore

$$F(I_i) = \int_{\ell_G \ln(I_i/I_0)}^{\infty} \frac{\exp(-l/\langle l \rangle)}{\langle l \rangle} dl, \tag{3.3}$$

so that the PDF for the emission signal I_i of such photon is [6]

$$P(I_i) = \frac{A}{I_i^{\varphi}}, \tag{3.4}$$

with $I_i \geq I_0$, A as a normalization constant, and the power-law exponent [6]

$$\varphi(\lambda) = 1 + \frac{\ell_G}{\langle l \rangle}. \tag{3.5}$$

So, the total emission intensity due to a large number N_{ph} of exiting photons with wavelength λ_n is given by the following sum of random variables,

$$I_n = \sum_{i=1}^{N_{\text{ph}}} I_i(\lambda_n). \tag{3.6}$$

At this point, we observe that the statistical properties of the sum of a large number of random variables are governed either by the central limit theorem (CLT) or by the generalized CLT [20, 21]. Essentially, we notice that if the identically distributed random variables are uncorrelated (or even if they present finite-extent correlations), and if the second moment of their PDF is finite, then the CLT assures that the PDF of their sum converges to a Gaussian as the number of the terms in the sum progressively increases, with the fluctuations driven by the Brownian (normal) dynamics [20, 21]. On the other hand, if the second moment of the PDF of the random variables diverges, thus the generalized CLT states that their sum is asymptotically described by the family of Lévy α-stable distributions, with the Fourier transform to the k-space given by the characteristic function [20, 21]

$$\bar{P}(k) = \exp\{-|ck|^{\alpha}[1 - i\beta \ \text{sgn}(k)\Phi] + ikv\}. \tag{3.7}$$

The Lévy index $\alpha \in (0,2]$ is the most important parameter, since it drives the type of statistics that characterizes the fluctuations of the random variable. Indeed, whereas strong fluctuations, with relevant deviations from the Gaussian behavior, are associated with parameter values in the range $0 < \alpha < 2$, the Gaussian statistics with weak fluctuations and the result of the CLT are recovered for the limit value $\alpha = 2$. The other parameters describe the asymmetry or skewness of the distribution ($\beta \in [-1,1]$), location ($\nu \in (-\infty,\infty)$), and scale ($c \in (0, \infty)$), along with the function $\Phi(k) = -(2/\pi)\ln|k|$ if $\alpha = 1$, whereas $\Phi = \tan(\pi\alpha/2)$ if $\alpha \neq 1$.

The connection between the Lévy statistics of the sum of a large number of random variables and the power-law PDF of I_i, given by

Eq. (3.4), can be understood as follows. By considering the Lévy α-stable distribution $P(I_n)$ as the Fourier transform of Eq. (3.7), it is known that its large-I_n asymptotic behavior has the form of a power-law function with a heavy tail,

$$P(I_n) \sim \frac{A}{I_n^{\mu}} \tag{3.8}$$

where the exponent μ is related to the Lévy index α through [20, 21]

$$\mu = \alpha + 1, \quad 0 < \alpha < 2. \tag{3.9}$$

Conversely, it is also true [20, 21] that random variables with power-law distributions are governed by the Lévy PDF with $\alpha = \mu - 1$ if $1 < \mu < 3$ (diverging second moment), and by the $\alpha = 2$ Gaussian statistics if $\mu \geq 3$ (finite second moment). Therefore, Lepri and co-authors essentially conclude in Ref. [6] that if $1 < \varphi < 3$ in Eq. (3.4) then a Lévy statistical behavior of output intensities with strong fluctuations emerges in the RL system, whereas if $\varphi \geq 3$ a weakly fluctuating Gaussian regime takes place. Their findings are corroborated [6] by Monte Carlo numerical simulations.

3.3 Lévy Statistics in Random Lasers: The Langevin Dynamics Approach

An alternative theoretical description of the Lévy statistics of intensities in RLs was derived [22, 23] by our research group through a rather distinct approach based on the system of coupled Langevin dynamics equations for the amplitudes of the resonance modes, as described below.

We start by considering a disordered dielectric medium of a RL. Since the system may present a large number N of overlapping modes, the leading order electromagnetic field solution of the nonlinear Maxwell equations is [24–27]

$$\vec{E}(\vec{r}, t) = \mathrm{Re}\left\{\sum_{k=1}^{N} a_k(t)\vec{E}_k(\vec{r})\exp(-i\omega_k t)\right\}, \tag{3.10}$$

$$\vec{H}(\vec{r}, t) = \mathrm{Re}\left\{\sum_{k=1}^{N} a_k(t)\vec{H}_k(\vec{r})\exp(-i\omega_k t)\right\}, \tag{3.11}$$

in which the field amplitudes appear above modified by the complex time-dependent dimensionless prefactor $a_k(t)$, arising from the nonlinearity of the amplifying medium. We thus generally take that

$$a_k(t) = A_k(t)\exp(i\phi_k(t)), \tag{3.12}$$

with $A_k(t)$ and $\varphi_k(t)$ denoting, respectively, the amplitude and phase of mode k. The RL intensity signal of mode k is given by the time average

$$I_k = \langle I_k(t)\rangle, \tag{3.13}$$

where

$$I_k(t) = c_k[A_k(t)]^2 \tag{3.14}$$

and c_k as a time-independent prefactor related to the position-dependent part of the electromagnetic field amplitudes in Eqs. (3.10) and (3.11). The total intensity is thus written as

$$I = \sum_{k=1}^{N} I_k. \tag{3.15}$$

In a series of remarkable works by Angelani, Leuzzi, Conti, Antennucci, and collaborators, whose main results are reviewed in Chapter 5, a theoretical phase diagram for laser systems in random amplifying media was built in the last decade [24–33], based on the system of N coupled Langevin equations for the complex slow-amplitude modes $a_k(t)$,

$$\frac{da_k}{dt} = -\frac{\partial H}{\partial a_k^*} + F_k, \tag{3.16}$$

where F_k is a Gaussian (white) uncorrelated noise and the general complex-valued functional H is (following closely the notation of Ref. [31])

$$H = \sum_{\{k_1,k_2\}'} g^{(2)}_{k_1k_2} a_{k_1} a^*_{k_2} + \frac{1}{2} \sum_{\{k_1,k_2,k_3,k_4\}'} g^{(4)}_{k_1k_2k_3k_4} a_{k_1} a^*_{k_2} a_{k3} a^*_{k_4}. \tag{3.17}$$

The symbol $\{\ldots\}'$ in the functional H above implies the frequency-matching conditions [24–33],

$$|\omega_{k_1} - \omega_{k_2}| < \gamma, \quad |\omega_{k_1} - \omega_{k_2} + \omega_{k_3} - \omega_{k_4}| < \gamma, \tag{3.18}$$

in the quadratic and quartic terms, respectively, with γ denoting the finite linewidth of the modes. Importantly, this formulation has been shown [28, 29] to apply to RL systems operating in both closed and open cavities, the latter constituting the case of general RL systems.

The physical origin of the quadratic coupling in H lies in the spatially inhomogeneous refractive index [observe that the spatial randomness of the active medium implies a static refractive index with a spatially random profile $n(\vec{r}) = c(\mu_0\varepsilon(\vec{r}))^{1/2}$], as well as in the nonuniform distribution of the gain and effective damping contribution due to the eventual cavity leakage. In systems with null or weak leakage, with negligible off-diagonal contribution, the real part of the diagonal coupling are the coefficient rates of amplification (γ_k) and radiation loss (α_k) of mode k,

$$g_{kk}^{(2)R} = \mathrm{Re}\left\{g_{kk}^{(2)}\right\} = \alpha_k - \gamma_k. \tag{3.19}$$

On the other hand, the quartic coupling in Eq. (3.17) is related to the nonlinear $\chi^{(3)}$-susceptibility with random spatial profile [24–33].

By substituting Eqs. (3.12) and (3.14) into Eqs. (3.16) and (3.17),

$$\frac{1}{c_{k_2}}\frac{dI_{k_2}}{dt} = -2\mathrm{Re}\left\{\sum_{\{k_1\}'} g_{k_1k_2}^{(2)} a_{k_1} a_{k_2}^* \right.$$
$$\left. +\frac{1}{2}\sum_{\{k_1,k_3,k_4\}'}\left[g_{k_1k_2k_3k_4}^{(4)} + g_{k_1k_4k_3k_2}^{(4)}\right] a_{k_1} a_{k_2}^* a_{k_3} a_{k_4}^* + a_{k_2}^* F_{k_2}\right\}. \tag{3.20}$$

The sum indicated by the symbol $\{k_1, k_3, k_4\}'$ in the quartic coupling in Eq. (3.20) generally involves three classes of mode combinations [24, 34]:

$$\omega_{k_1} = \omega_{k_2} \text{ and } \omega_{k_3} = \omega_{k_4}, \tag{3.21}$$

$$\omega_{k_1} = \omega_{k_4} \text{ and } \omega_{k_2} = \omega_{k_3}, \tag{3.22}$$

and the remaining possibilities satisfying the frequency-matching conditions, which have been usually disregarded [24, 34].

We now comment on the optical random noise in Eq. (3.20). For input excitation powers below the RL threshold, it is expected that the stochastic noise is dominated by the contribution from the spontaneous emissions, being therefore essentially independent of the field amplitudes. This constitutes the so-called additive noise

[35–37]. However, for higher input powers near and above the RL threshold the stimulated emissions start to play a relevant role, together with the increase of the nonlinear effects, so that the stochastic fluctuations should enhance with the field amplitudes, thus characterizing a multiplicative noise. In general, we consider that the optical noise in the RL system is a sum of additive and multiplicative statistically independent stochastic processes [35–37], which can be expressed by

$$F_k(t) = F_k^{(0)}(t) + a_k(t)\, F_k^{(1)}(t). \tag{3.23}$$

The absence of any kind of correlation in the white noise implies

$$\left\langle F_k^{(0)}(t) \right\rangle = 0 \text{ and } \left\langle F_k^{(0)}(t) F_{k'}^{(0)}(t') \right\rangle = 2q\delta_{k,k'}\delta(t-t'), \tag{3.24}$$

$$\left\langle F_k^{(1)}(t) \right\rangle = 0 \text{ and } \left\langle F_k^{(1)}(t) F_{k'}^{(1)}(t') \right\rangle = 2Q\delta_{k,k'}\delta(t-t'), \tag{3.25}$$

$$\left\langle F_k^{(0)}(t) F_{k'}^{(1)}(t') \right\rangle = 0, \tag{3.26}$$

in which $\delta_{k,k'}$ and $\delta(t-t')$ denote, respectively, the Kronecker and Dirac deltas, and the parameters q and Q control the respective magnitudes of the additive and multiplicative parts of the stochastic fluctuations. By applying Eqs. (3.12) and (3.14), we thus write

$$Re\left\{a_k^* F_k\right\} = \sqrt{\frac{I_k}{c_k}}\left[F_k^{(0)R}\cos(\phi_k) + F_k^{(0)I}\sin(\phi_k)\right] F_k^{(0)}(t) + \frac{I_k}{c_k} F_k^{(1)R}. \tag{3.27}$$

We now consider in Eq. (3.12) that the phases $\{\phi_k\}$ of the modes vary much more rapidly [24, 25] than the related amplitudes $\{A_k\}$, so that they can be averaged out, leading to the following Fokker–Planck equation [22, 23] for the probability density function (PDF) of output intensities $P(\{I_k\}, t)$

$$\frac{\partial P}{\partial t} = -\sum_k \frac{\partial}{\partial I_k}\left[\left(-d_k I_k - b_k I_k^2 + 2QI_k\right) P\right] + 2Q\sum_k \frac{\partial^2}{\partial I_k^2}\left(I_k^2 P\right), \tag{3.28}$$

where

$$b_k = \frac{2g_{kkkk}^{(4)R}}{c_k} \tag{3.29}$$

and, with the use of Eq. (3.19),

$$d_k = \sum\nolimits_{n \neq k} \left(g^{(4)R}_{kknn} + g^{(4)R}_{knnk} + g^{(4)R}_{nkkn} + g^{(4)R}_{nnkk} \right) \frac{I_n}{c_n} - 2\left(\gamma_k - \alpha_k\right). \tag{3.30}$$

We observe that, by averaging out the rapidly evolving phases in order to obtain Eq. (3.28), the so-called free running approximation has been employed, in which the dynamics of the phases and amplitudes are considered as eventually decoupled [24, 25]. In this context, some features of the original model might be possibly removed, such as the phase-locking regime predicted in [28, 29]. Also, in a statistical physics approach we observe that integrating out irrelevant degrees of freedom should be properly done in the partition function, rather than in the Hamiltonian level, though the technical difficulties involved are rather challenging.

In order to search for solutions of Eq. (3.28), we first comment on their general dynamics properties. By considering the ansatz

$$P(\{I_k\}, t) = P^{(0)}(\{I_k\}) \exp(-\nu t), \tag{3.31}$$

where

$$P^{(0)}(\{I_k\}) = P(\{I_k\}, t) \tag{3.32}$$

represents the initial condition, and the fact that the boundary condition related to the physical constraint of finite energy of the pump source must satisfy

$$P(\{I_k\} \to \infty, t) = 0, \tag{3.33}$$

then the solution of Eq. (3.28) can be expressed in terms of an eigenfunction expansion associated with the (either discrete or continuous) set of eigenvalues $\{-\nu_i\}$, so that

$$LP_i^{(0)}(\{I_k\}) = -\nu_i P_i^{(0)}(\{I_k\}), \tag{3.34}$$

in which the second-order differential Fokker–Planck operator is

$$L(\{I_k\}) = -\sum_k \frac{\partial}{\partial I_k} \left(-d_k I_k - b_k I_k^2 + 2QI_k\right) + 2Q \sum_k \frac{\partial^2}{\partial I_k^2} I_k^2. \tag{3.35}$$

In general, exact solutions of Fokker–Planck equations such as Eq. (3.28), or the associated eigenvalue equations, Eq. (3.34), can be hardly ever found [37], mainly in the case when L is nonlinear, which actually corresponds to the present context. In spite of this, some considerations based on the physical characteristics of the actual emission spectra can be applied in order to allow the analytical treatment of Eq. (3.28).

We are interested in the time-independent stationary properties of the distribution that emerges as the steady-state solution of Eq. (3.28), that is,

$$\frac{\partial P}{\partial t} = 0. \tag{3.36}$$

At this point, we observe from the experimental results in 1D as well as in 3D RL systems [1–19] that the overall behavior of the mode intensities $\{I_k\}$ depends strongly on the excitation power regime. Actually, three different types of experimental behaviors emerge, as follows.

First, in the regime below the random laser threshold, the fluctuations in the intensity values I_k of any mode k, from shot to shot in a pulsed laser or in time intervals of the integration time in a cw laser source, are rather weak. Indeed, this behavior corresponds to the Gaussian statistics of fluctuations, as commented above. In this case, the variance around the mean of the distribution of values of I_k is relatively small. Secondly, as the RL threshold is approached, a strong linewidth reduction takes place. As a consequence, the strong fluctuations become essentially restricted to the modes around the one with maximum output intensity, whereas the remaining modes still present weakly fluctuating small output intensity. This maximum intensity in this regime has been characterized experimentally by a Lévy-like statistical behavior, with heavy-tail power-law decay of the PDF of intensities in the large-I asymptotical regime. Lastly, well above the threshold another Gaussian regime emerges, which is, however, very distinct from the first prelasing Gaussian one. Indeed, this RL regime at high excitation powers has been shown to display self-averaging of the gain [11, 12], related to a more equitable distribution of the gain among the lasing modes due to the increasing in the strength of the mode interactions, along with the somewhat reduced intensity fluctuations (of intermediate magnitude between the prelasing

Gaussian and the threshold Lévy regimes) and the strong linewidth reduction.

Now, by considering in Eq. (3.20) that the contribution of the off-diagonal terms is negligible and also that the number of modes is sufficiently large, so that the sum in Eq. (3.30) approaches an effective value, then the Langevin equation for the dynamics of the output intensity of mode k implies the approximate Fokker–Planck equation for $P(I_k, t)$,

$$\frac{\partial P}{\partial t} = -\frac{\partial}{\partial I_k}\left[\left(-d_k I_k - b_k I_k^2 + 2QI_k\right)P\right] + 2Q\frac{\partial^2}{\partial I_k^2}\left(I_k^2 P\right). \quad (3.37)$$

The steady-state solution of Eq. (3.37) can be found by straightforward integration, leading to [22, 23]

$$P(I_k) = \frac{A_k}{I_k^{\mu_k}} \exp\left(-b_k I_k/2Q\right), \quad (3.38)$$

with $I_k > 0$, A_k as a normalization constant, and the exponent given by

$$\mu_k = 1 + \frac{d_k}{2Q}. \quad (3.39)$$

It is interesting to compare the results for the PDF of intensities derived from the approaches considering the stochasticity of the photons trajectories, Eqs. (3.4) and (3.5), with Eqs. (3.38) and (3.39) above, which emerge in the Langevin dynamics context. We first notice that the PDF of intensities in the latter case presents an exponential attenuation to the heavy power-law behavior, which is absent in the former approach and is important from a realistic point of view (see discussion below). Further, we also observe a noteworthy similarity between the power-law exponents, Eqs. (3.5) and (3.39), that may reveal some possible connection between the ratio of the gain length of the active medium ℓ_G to the average length of the photon paths $\langle l \rangle$ and the ratio of the parameter d_k, expressing the nonlinear coupling between the modes in the disordered medium, as well as the coefficients of amplification and radiation losses, to the magnitude of the multiplicative optical white noise present in the sample.

We also observe that an alternative approach to the statistics of output intensities in the prelasing regime is as follows. For input powers below the RL threshold, the gain coefficient is relatively low, if compared to the losses, i.e., $\gamma_k < \alpha_k$. Moreover, as mentioned above, in this prelasing regime the stochastic noise in Eq. (3.23) is purely additive. These facts considered together in the Langevin dynamics and associated Fokker–Planck give rise to a Gaussian distribution of output intensities below the RL threshold.

It is also worth commenting on the physical connection between the findings reported in Eqs. (3.38) and (3.39) and the parameters which are relevant in the experimental context, such as the input excitation energy and the disorder strength associated with the random medium of the RL system. Indeed, in the following we shall illustrate this point with the example of a RL system. To this end, we start by writing the displacement and magnetic induction vectors, respectively,

$$\vec{D} = \varepsilon_0 n^2 \vec{E} + \vec{P}_{NL} \text{ and } \vec{B} = \mu_0 \vec{H}, \tag{3.40}$$

in which the electric $\vec{E}$ and magnetic $\vec{H}$ fields are given by Eqs. (3.10) and (3.11), respectively, ε_0 and μ_0 denote the respective electrical permissivity and magnetic permeability in the vacuum, and we also assume that the medium has null magnetization. With these considerations in mind, the nonlinear polarization vector is given in leading order by

$$\begin{aligned}\vec{P}_{NL}(\vec{r}, t) = \text{Re}\{&\Sigma_{\{k_1,k_2,k_3,k_4\}}, \Sigma_{\{\alpha_1,\alpha_2,\alpha_3,\alpha_4\}}\hat{e}_{\alpha_1} \\ &\chi^{(3)}_{\alpha_1\alpha_2\alpha_3\alpha_4}\left(\omega_{k_1}; \omega_{k_3}, \omega_{k_4}, -\omega_{k_2}, \vec{r}\right) \\ &E^{\alpha_2}_{k_2}(\vec{r})E^{\alpha_3}_{k_3}(\vec{r})E^{\alpha_4}_{k_4}(\vec{r}) \exp(-i\omega_{k_1} t)\},\end{aligned} \tag{3.41}$$

where $\alpha_i = x, y, z$ for $i = 1, 2, 3, 4$. On the other hand, in the case of a medium of volume V, the fourth-order tensor in Eq. (3.17) is given by [24]

$$\begin{aligned}g^{(4)}_{k_1k_2k_3k_4} = \frac{1}{2i}\int_V &dV\, \Sigma_{\{\alpha_1,\alpha_2,\alpha_3,\alpha_4\}}\chi^{(3)}_{\alpha_1\alpha_2\alpha_3\alpha_4}\left(\omega_{k_1}; \omega_{k_3}, \omega_{k_4}, -\omega_{k_2}, \vec{r}\right) \\ &\times E^{\alpha_2}_{k_2}(\vec{r})E^{\alpha_3}_{k_3}(\vec{r})E^{\alpha_4}_{k_4}(\vec{r}) \exp(-i\omega_{k_1} t).\end{aligned} \tag{3.42}$$

Consequently, a direct comparison between Eqs. (3.41) and (3.42) allows to write the contribution to the average electromagnetic energy arising from the nonlinear character of the active medium of the RL in the form

$$E_{NL} = \frac{1}{4}\sum_{n\neq k}\left(g^{(4)R}_{kknn} + g^{(4)R}_{knnk} + g^{(4)R}_{nkkn} + g^{(4)R}_{nnkk}\right)\frac{I_n}{c_n}\frac{I_k}{c_k}. \tag{3.43}$$

By comparing Eqs. (3.30) and (3.43), we infer that an increasing excitation power generally leads to a higher electromagnetic energy and a larger sum in these equations, contributing to raise the exponent μ_k in Eq. (3.39).

On the other hand, the noise amplitude Q is proportional to the heat-bath temperature of the system, and therefore plays an opposite role with respect to the input pump energy in Eq. (3.39), tending to decrease the value of μ_k. Indeed, it has been experimentally recognized that the increase in the excitation power or the decrease of the temperature promotes the same qualitative effect in RL systems [24, 25, 38, 39].

Moreover, the fourth-order tensor elements in Eqs. (3.39) and (3.43) also embody the random spatial disorder properties of the active medium, as can be seen, for example, from the explicit spatial dependence in the fields and nonlinear susceptibility in Eq. (3.42). In particular, the strength of the disorder can be measured through the variance of the distributions of values of the tensor elements. Indeed, in the absence of any sort of disorder the dispersion of these quantities is null, so that

$$g^{(4)}_{k_1k_2k_3k_4} = g. \tag{3.44}$$

In this case, the disorder-induced RL phase is not possible [24, 25].

We now analyze the statistical regimes of intensity fluctuations from the probability distribution $P(I_k)$ described in Eqs. (3.38) and (3.39). This analysis is actually similar to that performed right below Eqs. (3.8) and (3.9).

We initially recall the PDF of intensities expressed by Eq. (3.38) identifies an exponentially truncated Lévy-like distribution for $1 < \mu_k < 3$ and $0 < \alpha < 2$. In this case, strong fluctuations in I_k emerge, and the ultraslow convergence predicted by the CLT to the $\alpha = 2$ Gaussian behavior due to the presence of the exponential factor in Eq. (3.38) is achieved only in the rather long term [40–42] (see also discussion below). In contrast, for values $\mu_k \leq 1$ or $\mu_k \geq 3$ the CLT assures fast convergence to the $\alpha = 2$ Gaussian statistics of weakly fluctuating intensities. Therefore, for a given disorder strength in a sample, an increasing excitation power raises [22, 23] the value of μ_k, and the statistics of maximum emission intensities shifts progressively from an initial Gaussian ($\mu_k \leq 1, \alpha = 2$), to a Lévy-type ($1 < \mu_k < 3, 0 < \alpha < 2$), and to a subsequent Gaussian ($\mu_k \geq 3, \alpha = 2$) regime.

These results are summarized in Fig. 3.1, in which the experimental data on the erbium-based RFL system are also shown for comparison.

We observe at this point that a PDF with diverging second moment, as those attracted by the Lévy α-stable distributions due to the generalized CLT, actually represents an unphysical possibility in the present case of intensity measurements. Therefore, at first sight this fact would in principle prevent the experimental realization of the asymptotic connection described above regarding the power-law long-tailed PDFs, Eq. (3.8), with exponent $1 < \mu < 3$ (diverging variance) and Lévy distributions with index $0 < \alpha = \mu - 1 < 2$. Nevertheless, it has been demonstrated [40–42] that a *truncated* power-law PDF, with a large but finite second moment, behaves rather similarly to the Lévy PDF to a considerable extent, thus defining what sometimes is called in the literature as a Lévy-type (or Lévy-like) statistical regime. In this case, the crossover to the Gaussian dynamics predicted by the CLT is attained only in the very long term [40].

We also comment that theoretically justified truncation schemes have been suitably implemented, for example, by restricting the values of the random variable to a finite range [40–42] or by tempering the power law with an exponential attenuation [12], in a way very much alike the form described in Eq. (3.38). So, we conclude that the reports of Lévy distributions of intensities in RLs with index $0 < \alpha < 2$, experimental, numerical, or analytical, should be properly interpreted as representative of this extensive Lévy-like statistical regime of intensity measurements.

3.4 Contribution of Higher-Order Nonlinearities

We now discuss the contribution of higher-order nonlinearities to the results of the preceding section.

As the input power is increased, higher f-th-order optical nonlinear terms ($f > 4$) related to the nonlinear susceptibility

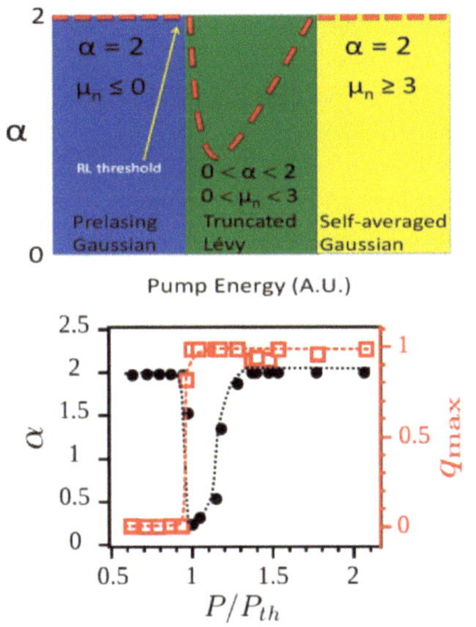

Figure 3.1 (Top) Schematic picture of the statistical regimes of the output intensity probability distribution in RL systems, which emerge from both theoretical approaches based on the stochastic trajectories of photons within the sample [6] and the Langevin dynamics equation [22, 23]. The parameter $\alpha \in (0,2]$ denotes the Lévy index of the family of Lévy α-stable distributions. As the excitation pump energy increases, the value of α shifts at the RL threshold from $\alpha = 2$ in the prelasing regime, in which weak Gaussian intensity fluctuations dominate, to a Lévy-like behavior with $0 < \alpha < 2$ nearly above the threshold, where strong intensity fluctuations set in. For higher input powers, a second $\alpha = 2$ Gaussian regime emerges with self-averaging of the gain, which is, however, very distinct from the prelasing regime. (Bottom) Experimental results for α (solid black circles, left axis) obtained for the erbium-based RFL [15]. P and P_{th} denote, respectively, the input excitation power and the RL threshold power. Notice the same progression of statistical regimes of the output intensity as the input energy is enhanced. A similar picture has been also found in the works by Uppu, Mujumdar and collaborators in 3D RL systems [7–12]. Empty red squares depict the values q_{max} of the Parisi overlap parameter for which its distribution is maximum. Observe that a transition from the prelasing replica-symmetric regime with $q_{\text{max}} \cong 0$ to the photonic RSB spin-glass phase with $q_{\text{max}} \cong 1$ also occurs at the RL threshold, simultaneously with the first shift from the Gaussian to the Lévy-like regime of intensity fluctuations. Such photonic spin-glass transition has been also demonstrated by Ghofraniha and collaborators in a 2D functionalized amorphous solid-state RL [33] (see, also, Chapter 6). Reproduced with permission from Ref. [15].

$\chi^{(f-1)}$ become progressively important to the PDF $P(I)$. More recently, these terms were considered [43, 44], leading to the so-called generalized Izrailev distribution of order f,

$$P(I) = N_f I^{\zeta_f - 1} \exp\left(-\sum_{k=2}^{f/2} c_k I^{k-1}\right), \tag{3.45}$$

where N_f is a normalization constant and the parameters ζ_f and c_k are associated with the nonlinear coefficients and noise amplitude.

We show in Fig. 3.2 the nice agreement with Eq. (3.45) using $f = 6$ of the experimental data of 10^5 spectra emitted by the RL system based on the Nd^{3+}-doped crystalline nanopowder for an excitation power $P/P_{\text{th}} = 3.9$ well above threshold, where higher order nonlinearities are relevant. For comparison, poorer fits are displayed with $f = 4$ and also using the predictions from random phase sum (RPS) based models, in which the electric field is assumed to be a sum of many complex individual contributions with random

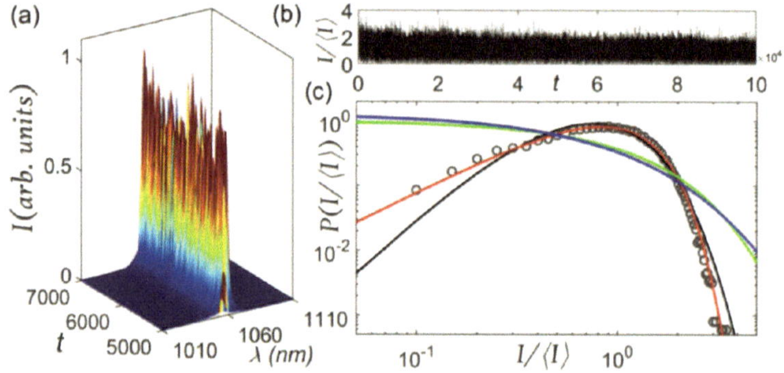

Figure 3.2 (a) A fraction of 2,000 out of the 10^5 spectra emitted by the RL system based on the Nd^{3+}-doped crystalline nanopowder for a relative excitation power $P/P_{\text{th}} = 3.9$ well above threshold. (b) Maximum intensities I (normalized by their average) as a function of the spectrum label $t = 1, 2, \ldots, 10^5$. (c) Circles depict the histogram of experimental intensities (in log-log plot), displaying an excellent fit of the experimental data to the Izrailev distribution of order $f = 6$ in the low-, intermediate-and high-I regimes [red line, Eq. (3.45) for $\zeta_6 = 2.6$, $c_2 = 0.60$, and $c_4 = 0.77$]. Poorer fits are also shown to the Izrailev distribution of order $f = 4$ [black line, with $\zeta_4 = 3.9$ and $c_2 = 3.8$], and the PDFs resulting from the RPS based models: exponential [green line] and K distributions [blue line]. Reproduced with permission from Ref. [44].

phases and amplitudes that can be either uncorrelated or weakly correlated (K distribution in the latter case and exponential in the former, with a Rayleigh PDF for the electric field).

3.5 Extreme-Value Analysis of Emitted Intensities

The fact discussed above that the PDF of emitted intensities is of Lévy-type, generally with large second moment and rather strong fluctuations, leads to the rare but non-negligible phenomenon of extreme values [45] (for further details, see also Chapter 4 of this book). In particular, this has been also studied in a RFL in Ref. [17], as briefly reviewed below.

The extreme-value analysis starts by subdividing the sequence of maximum intensity of a large (N) number of spectra into M blocks of equal size. By picking the largest intensity of each block, a new sequence of M intensity values is generated. According to the extreme value statistics (EVS) theory, the cumulative density function (CDF) of such maximum values has the form of a generalized extreme-value (GEV) distribution [45],

$$F(I) = \exp\left\{-\left[1+\xi\frac{(I-m)}{\sigma}\right]^{-1/\xi}\right\}, \tag{3.46}$$

where $1+\xi(I-m)/\sigma > 0$, and ξ, m and $\sigma > 0$ are, respectively, the shape, location, and scale parameters. The EVS theory also states that the GEV distributions can be cast into three classes [17]: (i) Weibull for $\xi < 0$ and I limited to an upper value; (ii) Gumbel for $\xi \to 0$, and (iii) Fréchet for $\xi > 0$.

A formal theoretical connection can be established [17] between the primary PDF $P(I)$ comprising all intensity values (i.e., not only the maxima of the blocks) and the associated asymptotic GEV distribution $F(I)$ of the maximum intensity values. Indeed, if the PDF $P(I)$ falls off asymptotically at large I as a power-law heavy tail, $P(I) \sim I^{-\mu}$, with $\mu > 1$, then its corresponding GEV distribution converges in the limit $N/M \to \infty$ to the Fréchet density function. Conversely, if $P(I)$ falls off faster than a power law, the convergence function is the $\xi \to 0$ Gumbel CDF.

From the discussion above, and in the context of the analysis of experimental data, we conclude that if the experimental CDF $F(I)$ best fits the Fréchet ($\xi > 0$) GEV distribution, a large-I power-law dependence can be inferred for the PDF $P(I)$, with exponent $\mu = 1 + 1/\xi > 1$. In this case, since the Lévy (Gaussian) regime is characterized by $1 < \mu < 3$ ($\mu \geq 3$), then we can define a parameter α' as a further experimental indicator of the statistics of intensity fluctuations, which is determined from the EVS analysis of maximum intensities, instead of from the direct analysis of $P(I)$ itself. In this context, we observe that $\alpha' = \mu - 1 = 1/\xi < 2$ if $\xi > 1/2$ (Lévy regime), and $\alpha' = 2$ if $0 < \xi \leq 1/2$ (Gaussian regime). The Fréchet boundary limit $\xi = 1/2$ thus separates the domains with Lévy and Gaussian behaviors. In contrast, if the best fit of the experimental $F(I)$ matches the $\xi \rightarrow 0$ Gumbel distribution, then the $\alpha' = 2$ Gaussian regime is implied. Of course, one would actually expect that $\alpha' = \alpha$ in the limit $N/M \rightarrow \infty$, thus indicating that the same statistics of intensity values should be obtained in this limit either from the EVS analysis of the CDF $F(I)$ (α') or by investigating the PDF $P(I)$ (α). Nevertheless, in real experimental studies this precise equivalence can be hindered, and one can possibly find $\alpha' \approx \alpha$ from experimental data.

Figure 3.3 shows the CDF $F(I)$ for various input powers for the erbium-based RFL [17]. A Fréchet CDF is identified just above the threshold, consistently with the Lévy statistical behavior of intensity fluctuations. Gaussian regimes are found well below and well above the threshold, characterized by distributions with $\xi \approx 0$, either Fréchet or Gumbel, and just below the threshold the strong increase of the shape parameter toward the $\xi = 1/2$ Fréchet limit signalizes the onset of the transition from the Gaussian to Lévy behavior.

3.6 Analytical Connection with the Photonic Spin-Glass Random-Laser Behavior

We finally discuss the connection of the above findings with the photonic spin-glass behavior and replica symmetry breaking (RSB) [16–19, 23–33,46–49]. For a comprehensive review of these works,

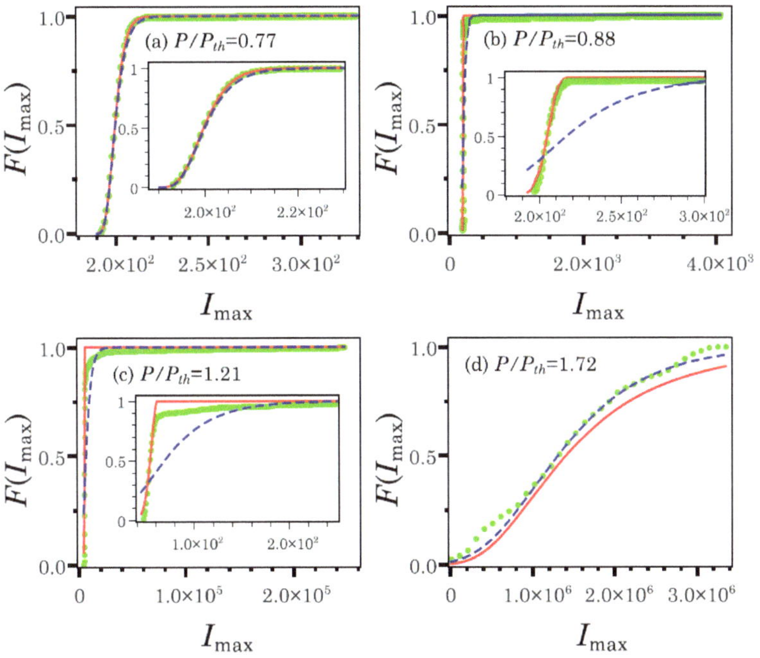

Figure 3.3 CDF $F(I)$ of maximum intensity values of the $M = 800$ blocks of spectra emitted by the erbium-based RFL system at (a) $P/P_{\text{th}} = 0.77$, (b) $P/P_{\text{th}} = 0.88$, (c) $P/P_{\text{th}} = 1.21$, and (d) $P/P_{\text{th}} = 1.72$. Experimental results from the analysis of the sequences of maximum block intensities are depicted in green circles. Best-fit curves to the GEV and Gumbel distributions are indicated by solid red and dashed blue lines, respectively. A Fréchet CDF is identified just above the threshold [(c) $\xi = 0.62 > 1/2$], consistently with the Lévy statistical behavior of intensity fluctuations. Gaussian regimes are found well below and well above the threshold, characterized by distributions with $\xi \approx 0$, either Fréchet [(a) $\xi = 0.08$] or Gumbel [(a) and (d): $\xi \to 0$]. Just below the threshold [(b) $\xi = 0.35$], the strong increase of the shape parameter toward the $\xi = 1/2$ Fréchet limit signalizes the onset of the transition from the Gaussian to Lévy behavior. The insets show details of the main plots. Reproduced with permission from Ref. [17].

both on the theoretical and experimental points of view, we refer the reader to Chapters 5 and 6 of this book.

We remark that, from the analytical side, the complex correspondence between the changes in the statistical behavior of intensity fluctuations in RL systems, discussed in the previous sections, and the RSB transition to the photonic spin-glass phase can be explained within the same theoretical framework. Indeed, the very same set of Langevin equations describing the dynamics of the complex slow-amplitude modes $\{a_k\}$, given by Eqs. (3.16) and (3.17), also provides the theoretical basis for the emergence of an RSB glassy phase in the RL system, among other photonic behaviors in the phase diagram.

To contextualize our discussion, as reviewed in Chapter 6 the characterization of a photonic RSB spin-glass phase in the RL regime requires the calculation of mode correlations either among amplitudes, phases or intensity fluctuations. In the latter, which has the advantage to be experimentally accessible, the overlap parameter is defined in Ref. [33] as

$$q_{\gamma\beta} = \frac{\sum_k \Delta_\gamma(k)\Delta_\beta(k)}{\sqrt{\left[\sum_k \Delta_\gamma^2(k)\right]\left[\sum_k \Delta_\beta^2(k)\right]}}, \tag{3.47}$$

where $\gamma, \beta = 1, 2, \ldots, N_s$ denote the replica labels, the average intensity at the wavelength indexed by k reads $\langle I \rangle (k) = \sum_{\gamma=1}^{N_s} I_\gamma (k)/N_s$, and the intensity fluctuation is $\Delta_\gamma (k) = I_\gamma (k) - \langle I \rangle (k)$. In this context, each output spectrum is considered a replica, i.e., a copy of the RL system under fairly identical experimental conditions. The PDF P(q), analogue to the Parisi order parameter in the RSB spin-glass theory of strongly disordered magnetic systems [50], describes the probability distribution of replica overlaps $q = q_{\gamma\beta}$, signaling a replica-symmetric paramagnetic regime or a RSB spin-glass phase if its maximum occurs exclusively at $q_{\max} = 0$ (no RSB) or also at values $|q_{\max}| \neq 0$ (RSB), respectively.

The bottom plot of Fig. 3.1 summarizes the connection between the photonic replica symmetric and RSB glassy phases and the distinct statistical regimes of intensity fluctuations. The right axis of the bottom plot of Fig. 3.1 displays the value of $q_{\max}$ as a function of the relative input power P/P_{th}. We first notice a remarkable agreement between the onset of the Lévy statistical regime of intensity fluctuations and the emergence of the photonic RSB glassy RL phase. So, besides signaling the Gaussian-to-Lévy shift, the RL threshold also marks the phase transition from the $q_{max} \cong 0$ replica-symmetric prelasing regime to the $q_{max} \cong 1$ spin-glass RL phase with RSB. This coinciding behavior, firstly reported in a 3D RL system [23], is thus also shared by the 1D erbium-doped RFL and other RLs.

We end this section by commenting that a complete theoretical understanding of these findings is still lacking [31]. Nevertheless, it is possible [22, 23] to trace back their common physical origin to the Langevin equations for the amplitudes of the modes, which, as discussed above, form the basis to explain both the distinct statistical regimes of intensity fluctuations and the RSB spin-glass behavior of RL systems.

3.7 Conclusions

The recent exploitation of random lasers (RLs) and random fiber lasers (RFLs) as photonic platforms for studying complex systems phenomena, as nontrivial Lévy-type statistics [1–19, 22, 23], photonic spin-glass properties [24–33, 43–46], turbulent-like behaviors [49–55], and other complex phases [56–59], has opened new cross disciplinary avenues for the understanding of the physical mechanisms underlying the RL emission properties.

In this chapter, we have reviewed two theoretical approaches that are able to provide theoretical explanations to two of the most remarkable properties of RLs and RFLs. In the first case, strong RL intensity fluctuations present heavy-tailed Lévy-like distributions, which contrast markedly with the Gaussian statistics of the prelasing regime. Second, the presence of strong disorder may hamper the synchronous oscillation of modes, thus producing a photonic replica-symmetry-breaking (RSB) spin-glass phase. The influence of high-order nonlinearities to the intensity fluctuations and the statistics of extreme events in RLs has been also considered here.

Interestingly, there seems to exist some universality component in those Gaussian-to-Lévy and paramagnetic-to-spin-glass photonic phase transitions. We thus hope that further experimental and theoretical works can be stimulated to tight the physical connections and to better explain the mechanisms of these complex photonic behaviors in RLs and RFLs.

Acknowledgments

The authors would like to thank the financial support from the Brazilian agencies Conselho Nacional de Desenvolvimento Científico e Tecnológico (CNPq), Coordenação de Aperfeiçoamento de Pessoal de Nível Superior (CAPES), Fundação de Amparo à Ciência e Tecnologia do Estado de Pernambuco (FACEPE), and Instituto Nacional de Fotônica (INFO).

References

1. Anglos, D., Stassinopoulos, A., Das, R. N., Zacharakis, G., Psyllaki, M., Jakubiak, R., Vaia, R. A., Giannelis, E. P., and Anastasiadis, S. H. (2004). Random laser action in organic-inorganic nanocomposites. *J. Opt. Soc. Am. B*, **21**, 208–213.
2. Sharma, D., Ramachandran, H., and Kumar, N. (2006). Lévy statistical fluctuations from a random aplifying medium. *Fluct. Noise Lett.*, **6**, L95–L101.
3. Sharma, D., Ramachandran, H., and Kumar, N. (2006). Lévy statistics of emission from a novel random amplifiying medium: an ooptical realization of the Arrhenius cascade. *Opt. Lett.*, **31**, 1806–1808.
4. Van der Molen, K. L., Mosk, A. P., and Lagendijk, A. (2006). Intrinsic intensity fluctuations in random lasers. *Phys. Rev. A*, **74**, 053808.
5. Mujumdar, S., Turck, V., Torre, R., and Wiersma, D. S. (2007). Chaotic behavior of a random laser with static disorder. *Phys. Rev. A*, **76**, 033807.
6. Lepri, S., Cavalieri, S., Oppo, G.-L., and Wiersma, D. S. (2007). Statistical regimes of random laser fluctuations. *Phys. Rev. A*, **75**, 063820.
7. Uppu, R., and Mujumdar, S. (2010). Statistical fluctuations of coherent and incoherent intensity in random lasers with nonresonant feedback. *Opt. Lett.*, **35**, 2831–2833.
8. Uppu, R., and Mujumdar, S. (2011). On the coherent modes of ultranarrowband random lasers with nonresonant feedback. *Appl. Opt.*, **50**, E13–E19.
9. Uppu, R., Tiwari, A. K., and Mujumdar, S. (2012). Identification of statistical regimes and crossovers in coherent random laser emission. *Opt. Lett.*, **37**, 662–664.

10. Uppu, R., and Mujumdar, S. (2013). Dependence of the Gaussian-Lévy transition on the disorder strength in random lasers. *Phys. Rev. A*, **87**, 013822.

11. Uppu, R., and Mujumdar, S. (2014). Lévy exponent as universal identifiers of threshold and criticality in random lasers. *Phys. Rev. A*, **90**, 025801.

12. Uppu, R., and Mujumdar, S. (2015). Exponentially tempered Lévy sums in random lasers. *Phys. Rev. Lett.*, **114**, 183903.

13. Ignesti, E., Tommasi, F., Fini, L., Lepri, S., Radhalakshmi, V., Wiersma, D., and Cavalieri, S. (2013). Experimental and theoretical investigation of statistical regimes in random laser emission. *Phys. Rev. A*, **88**, 033820.

14. Tommasi, F., Fini, L., Ignesti, E., Lepri, S., Martelli, F., and Cavalieri, S. (2018). Statistical outliers in random laser emission. *Phys. Rev. A*, **98**, 053816.

15. Lima, B. C., Gomes, A. S. L., Pincheira, P. I. R., Moura, A. L., Gagné, M., Raposo, E. P., de Araújo, C. B., and Kashyap, R. (2017). Observation of Lévy statistics in one-dimensional erbium-based random fiber laser. *J. Opt. Soc. Am. B*, **34**, 293–299.

16. De Araújo, C. B., Gomes, A. S. L., and Raposo, E. P. (2017). Lévy statistics and glassy behavior of light in random fiber lasers. *Appl. Sci.*, **7**, 644.

17. Lima, B. C., Pincheira, P. I. R., Raposo, E. P., Menezes, L. de S., de Araújo, C. B., Gomes, A. S. L., and Kashyap, R. (2017). Extreme-value statistics of intensities in a CW-pumped random fiber laser. *Phys. Rev. A*, **96**, 013834.

18. Gomes, A. S. L., Lima, B. C., Pincheira, P. I. R., Moura, A. L., Gagné, M., Kashyap, R., Raposo, E. P., and de Araújo, C. B. (2016). Glassy behavior in a one-dimensional continuous-wave erbium-doped random fiber laser. *Phys. Rev. A*, **94**, 011801(R).

19. Raposo, E. P., Gomes, A. S. L., Moura, A. L., Fewo, S. I., Pincheira, P. I. R., Lima, B. C., Carreño, S. J. M., Silva, A. F., Maia, L. J. Q., Jerez, V., Gagné, M., Kashyap, R., and Araújo, C. B. (2016). Universal phase transitions in random lasers. *Opt. Photon. News*, **27** (12), 34.

20. Samorodnitsky, G., and Taqqu, M. S. (1994). *Stable Non-Gaussian Random Processes,* Chapman and Hall, New York.

21. Uchaikin, V. V., and Zolotarev, V. M. (1999). *Chance and Stability: Stable Distributions and Their Applications*, VSP BV, Utrecht.

22. Raposo, E. P., and Gomes, A. S. L. (2015). Analytical solution for the Lévy-like steady-state distribution of intensities in random lasers. *Phys. Rev. A*, **91**, 043827.

23. Gomes, A. S. L., Raposo, E. P., Moura, A. L., Fewo, S. I., Pincheira, P. I. R., Jerez, V., Maia, L. J. Q., and de Araújo, C. B. (2016). Observation of Lévy distribution and replica symmetry breaking in random lasers from a single set of measurements. *Sci. Rep.*, **6**, 27987.
24. Angelani, L., Conti, C., Ruocco, G., and Zamponi, F. (2006). Glassy behavior of light. *Phys. Rev. Lett.*, **96**, 065702.
25. Angelani, L., Conti, C., Ruocco, G., and Zamponi, F. (2006). Glassy behavior of light in random lasers. *Phys. Rev. B*, **74**, 104207.
26. Leuzzi, L., Conti, C., Folli, V., Angelani, L., and Ruocco, G. (2009). Phase diagram and complexity of mode-locked lasers: from order to disorder. *Phys. Rev. Lett.*, **102**, 083901.
27. Conti, C., and Leuzzi, L. (2011). Complexity of waves in nonlinear disordered media. *Phys. Rev. B*, **83**, 134204.
28. Antenucci, F., Conti, C., Crisanti, A., and Leuzzi, L. (2015). General phase diagram of multimodal ordered and disordered lasers in closed and open cavities. *Phys. Rev. Lett.*, **114**, 043901.
29. Antenucci, F., Crisanti, A., and Leuzzi, L. (2015). Complex spherical 2 + 4 spin glass: a model for nonlinear optics in random media. *Phys. Rev. A*, **91**, 053816.
30. Antenucci, F., Crisanti, A., and Leuzzi, L. (2015). The glassy random laser: replica symmetry breaking in the intensity fluctuations of emission spectra. *Sci. Rep.*, **5**, 16792.
31. Antenucci, F., Crisanti, A., Ibañez-Berganza, M., Marruzzo, A., and Leuzzi, L. (2016). Statistical mechanics models for multimode lasers and random lasers. *Philos. Mag.*, **96**, 704–731.
32. Tommasi, F., Ignesti, E., Lepri, S., and Cavalieri, S. (2016). Robustness of replica symmetry breaking phenomenology in random laser. *Sci. Rep.*, **6**, 37113.
33. Ghofraniha, N., Viola, I., Di Maria, F., Barbarella, G., Gigli, G., Leuzzi, L., and Conti, C. (2015). Experimental evidence of replica symmetry breaking in random lasers. *Nat. Commun.*, **6**, 6058.
34. O'Bryan, I. C. L., and Sargent, I. M. (1973). Theory of multimode laser operation. *Phys. Rev. A*, **8**, 3071–3092.
35. Coffey, W. T., Kalmykov, Y. P., and Waldron, J. T. (2004) *The Langevin Equation: With Applications to Stochastic Problems in Physics, Chemistry, and Electrical Engineering*, 2nd ed., World Scientific, Singapore.
36. Pavliotis, G. A. (2014). *Stochastic Processes and Applications: Diffusion Processes, the Fokker-Planck and Langevin Equations*, Springer, New York.

37. Schenzle, A., and Brand, H. (1979). Multiplicative stochastic processes in statistical physics. *Phys. Rev. A*, **20**, 1628–1647.
38. Wiersma, D. S., and Cavalieri, S. (2001). A temperature-tunable random laser. *Nature*, **414**, 708–709.
39. Leonetti, M., and Conti, C. (2010). Haus/Gross-Pitaevskii equation for random lasers. *J. Opt. Soc. Am. B*, **27**, 1446–1451.
40. Mantegna, R. N., and Stanley, H. E. (1994). Stochastic process with ultraslow convergence to a Gaussian: the truncated Lévy flight. *Phys. Rev. Lett.*, **73**, 2946–2949.
41. Koponen, I. (1995). Analytic approach to the problem of convergence of truncated Lévy flights towards the Gaussian stochastic process. *Phys. Rev. E*, **52**, 1197–1199.
42. Bartumeus, F., Raposo, E. P., Viswanathan, G. M., and da Luz, M. G. E. (2014). Stochastic optimal foraging: tuning intensive and extensive dynamics in random searches. *PLoS One*, **9**, e106373.
43. González, I. R. R., Raposo, E. P., Carreño, S. J., Macêdo, A. M. S., Maldonado, M., Menezes, L. de S., Gomes, A. S. L., and de Araújo, C. B. (2020). Influence of fifth-order nonlinearities on the statistical fluctuations in emission intensities in a photonic open-cavity complex system. *Phys. Rev. A*, **102**, 063515.
44. González, I. R. R., Pincheira, P. I. R., Macêdo, A. M. S., Menezes, L. de S., Gomes, A. S. L., and Raposo, E. P. (2021). Intensity distribution in random lasers: comparison between a stochastic differential model of interacting modes and random phase sum based models. J. Opt. Soc. Am. B, to appear.
45. Sornette, D. (2006). *Critical Phenomena in Natural Sciences*, Springer, Berlin.
46. Pincheira, P. I. R., Silva, A. F., Carreño, S. J. M., Moura, A. L., Fewo, S. I., Raposo, E. P., Gomes, A. S. L., and de Araújo, C. B. (2016). Observation of photonic paramagnetic to spin-glass transition in specially-designed TiO_2 particles-based dye-colloidal random laser. *Opt. Lett.*, **41**, 3459–3462.
47. Moura, A. L., Pincheira, P. I. R., Reyna, A. S., Raposo, E. P., Gomes, A. S. L., and de Araújo, C. B. (2017). Replica symmetry breaking in the photonic ferromagnetic like spontaneous mode-locking phase of a multimode Nd:YAG laser. *Phys. Rev. Lett.*, **119**, 163902.
48. Basak, S., Blanco, A., and López, C. (2016). Large fluctuations at the lasing threshold of solid- and liquid-state dye lasers. *Sci. Rep.*, **6**, 32134.

49. Pierangeli, D., Tavani, A., Di Mei, F., Agranat, A. J., Conti, C., and DelRe, E. (2017). Observation of replica symmetry breaking in disordered nonlinear wave propagation. *Nat. Commun.*, **8**, 1501.
50. Mézard, M., Parisi, G., and Virasoro, M. A. (1987) *Spin Glass Theory and Beyond,* World Scientific, Singapore.
51. Crisanti, A., and Sommers, H.-J. (1992). The spherical p-spin interaction spin glass model: the statics. *Z. Phys. B*, **87**, 341–354.
52. Turitsyna, E. G., Smirnov, S. V., Sugavanam, S., Tarasov, N., Shu, X., Babin, S. A., Podivilov, E. V., Churkin, D. V., Falkovich, G., and Turitsyn, S. K. (2013). The laminar-turbulent transition in a fibre laser. *Nat. Photon.*, **7**, 783–786.
53. Sugavanam, S., Tarasov, N., Wabnitz, S., and Churkin, D. V. (2015). Ginzburg-Landau turbulence in quasi-CW Raman fiber lasers. *Laser Photon. Rev.*, **9**, L35–L39.
54. Walcsak, P., Randoux, S., and Suret, P. (2015). Statistics of a turbulent Raman fiber laser. *Opt. Lett.*, **40**, 3101–3104.
55. Churkin, D. V., Kolokolov, I. V., Podivilov, E. V., Vatnik, I. D., Nikulin, M. A., Vergeles, S. S., Terekhov, I. S., Lebedev, V. V., Falkovich, G., Babin, S. A., and Turitsyn, S. K. (2015). Wave kinetics of random fibre lasers. *Nat. Commun.*, **2**, 6214.
56. González, I. R. R., Lima, B. C., Pincheira, P. I. R., Brum, A. A., Macêdo, A. M. S., Vasconcelos, G. L., Menezes, L. de S., Raposo, E. P., Gomes, A. S. L., and Kashyap, R. (2017). Turbulence hierarchy in a random fibre laser. *Nat. Commun.*, **8**, 15731.
57. González, I. R. R., Raposo, E. P., Macêdo, A. M. S., Menezes, L. de S., and Gomes, A. S. L. (2018). Coexistence of turbulence-like and glassy behaviours in a photonic system. *Sci. Rep.*, **8**, 17046.
58. Macêdo, A. M. S., González, I. R. R., Raposo, E. P., Menezes, L. de S., and Gomes, A. S. L. (2019). Turbulent intermittency in a random fiber faser. *Atoms*, **7**, 43.
59. Raposo, E. P., González, I. R. R., Macêdo, A. M. S., Lima, B. C., Kashyap, R., Menezes, L. de S., and Gomes, A. S. L. (2019). Evidence of Floquet phase in a photonic system. *Phys. Rev. Lett.*, **122**, 143903.

Chapter 4

Lévy Fluctuations of Intensity in Random Lasers

Sushil Mujumdar and Ravitej Uppu
Nano-optics and Mesoscopic Optics Laboratory, Tata Institute of Fundamental Research, Mumbai 400005 India
mujumdar@tifr.res.in

4.1 Introduction

Random lasers are unique optical devices that aim to obtain laser-like emission upon exploiting the structural randomness of the amplifying medium. In a way, they are antipodal to conventional lasers, which rely on sophisticated engineering to obtain highly coherent emissions. A conventional laser essentially relies on three components, namely, a gain medium that provides the necessary amplification upon excitation, an energy pump that excites the gain medium to create a population inversion, and a high-finesse resonator, which assists in increasing the interaction time of the light with the inverted medium. The resonator, in its commonest form, is effectively a combination of two mirrors facing each other, which creates a standing wave pattern for select wavelengths of light

Lévy Statistics and Spin Glass Behavior in Random Lasers
Edited by Anderson S. L. Gomes, André L. Moura, Cid B. de Araújo, and Ernesto P. Raposo

ISBN 978-981-4968-55-3 (Hardcover), 978-1-003-33618-1 (eBook)
www.jennystanford.com

determined by the condition $2L_{cav} = mn_{ref}\lambda$, where L_{cav} is the cavity length, m is the mode order, and n_{ref} is the intracavity refractive index. These light waves are longest-lived inside the resonator, and experience maximal gain. A random laser is achieved by replacing the mirrors with alternative sources of feedback that have inherent randomness. For instance, an ensemble of small particles can create the necessary randomness, which can act as a feedback mechanism. A light wave undergoes multiple scattering in this ensemble, and after every scattering event, its propagation length within the amplifying medium increases. Thus, the random scattering acts as a path extender to the light, somewhat akin to what the mirrors in the conventional laser achieve. Naturally, the degree of feedback is determined by the strength of the disorder in the system.

In the particulate scattering example above, the disorder strength can be quantified by two length scales, namely, the mean free path and the system size L_{sys}. Two kinds of mean free paths are typically used for describing the disordered system; the scattering mean free path (ℓ_s), which is the average distance between two scattering events, or the transport mean free path (ℓ^*), which is the average distance the incident wave travels before its momentum gets randomized. Mathematically, $\ell_s = 1/N\sigma_s$, and $\ell^* = \frac{\ell_s}{1-\langle\cos\theta_s\rangle}$, where N is the scatterer number density, σ_s is the scattering cross-section of each scatterer, $\langle\cos\theta_s\rangle$ is the scattering anisotropy of the individual scatterer, or the average cosine of the scattering angle θ_s. Clearly, ℓ^* accounts for the anisotropy in scattering, which is often present in real samples. On the other hand, for point scatterers, $\ell_s = \ell^*$. In the multiple scattering regime, when $\lambda \ll \ell^*$, light transport in the medium can be modeled in the geometric optics approximation as a random walk of photons, essentially modeling the diffusion of light. For $\ell^* \lesssim \lambda$, the diffusion approximation breaks down as the interference effects in the propagation influence the light transport and may result in inhibition of transmission. This represents the *Anderson localized regime*, predicted for the transport of electrons through disordered metals (Anderson, 1958). In three dimensions, Anderson localization is extremely hard to achieve and transport is almost always diffusive.

The fact that multiple scattering increases the dwell time for light and hence can be used for amplification was theoretically proposed

(Ambartsumyan et al., 1966, 1970) using randomized Fabry-Perot cavities with roughened mirrors. Letokhov extended this concept for the creation of multiple scattering in a collection of point scatterers with gain in the diffusive regime (Letokhov, 1968), which was experimentally verified in Nd-doped powders (Markushev et al., 1986). The breakthrough demonstration of random lasing was reported in 1994 in a dye-scatterer system (Lawandy et al., 1994)[1]. A rapid bandwidth collapse in the emission along with a laser-like threshold was observed in a medium comprising a laser dye suspension of dielectric nanoparticles. The emission featured a stable spectrum with one or two peaks and a bandwidth of $\sim$4–5 nm, which is about an order of magnitude smaller than the fluorescence bandwidth of the dye (Lawandy et al., 1994; Balachandran and Lawandy, 1996; Mujumdar and Ramachandran, 2000). Importantly, this system separated the scattering (dielectric particles) and gain (laser dye) media, thus allowing independent control of these two parameters. These improvements reignited the research interest in these systems for understanding the intricacies to help in tailoring them for applications. Since these systems realized laser-like behavior through a synergy between amplification and multiple scattering off random sites, they were termed *random lasers* (Wiersma, 2008). Furthermore, as diffusion is the feedback mechanism in these weak scattering systems, they have also been called *diffusive random lasers*. Random lasing has since been demonstrated in a range of media, such as minute clusters of nanoparticles, cold atom clouds, optical fibers with lengths of the order of several tens of kilometers, human tissues, etc. (Cao et al., 2000; Turitsyn et al., 2010; Baudouin et al., 2013; Polson and Vardeny, 2004; Tulek et al., 2010; Tiwari and Mujumdar, 2013; Liu et al., 2014). Models for light undergoing diffusion along with amplification have been constructed using purely diffusive transport of intensity (John and Pang, 1996; Wiersma and Lagendijk, 1996; Berger et al., 1997; Mujumdar and Ramachandran, 2000; Florescu and John, 2004).

---

[1]We happily note that one of the co-authors of this seminal paper is the editor of this book.

4.1.1 Experimental Measurements

The random lasers used in our experiments were of the typical dye-scatterer type, consisting of the laser dye Rhodamine 6G dissolved in methanol (concentration of 2.5 mM) into which dielectric nanoparticles (ZnO, diameter $d \sim 20$ nm) were suspended. Modifying the volume fraction of the scatterers tuned the disorder strength of the random laser sample. In a static measurement, a Coherent Backscattering experiment was set up to measure the transport mean free path (ℓ^*), which quantified the disorder strength (Wolf and Maret, 1985; van Albada and Lagendijk, 1985). Statistical studies on the intensity fluctuations described in this chapter were carried out on systems with disorder strength ranging between $\ell^* = 200$ μm (~35 mg/ml) and $\ell^* = 1500$ μm (~4.5 mg/ml). Clearly, $\ell^* \gg \lambda$, these systems belong to the weak scattering regime, i.e., far from the localization regime. However, in the context of our investigations, systems with scattering strengths close to $\ell^* = 200$ μm are referred to as *strongly scattering* and those close to $\ell^* = 1500$ μm are referred to as *weakly scattering*. This definition is motivated by the stronger diffusion in the $\ell^* = 200$ μm systems, the consequences of which are evident in the statistics discussed later.

This section discusses the experiments performed on a random laser with $\ell^* \sim 1500$ μm. This choice of disorder strength for discussion is due to the distinct spectral line-shape regimes present in the diffusive and coherent random lasers. A volume of 1 ml of the suspension was taken in a quartz cuvette, with a magnetic stirrer placed inside to agitate the suspension in order to prevent the sedimentation of the nanoparticles.

Figure 4.1 shows the schematic of the experimental setup employed for observing and analyzing the random laser emission. The sample was pumped using the second harmonic ($\lambda = 532$ nm) of a pulsed Nd:YAG laser. To observe coherent random lasing, a 30 ps pulse width laser was used. In situations requiring measurements of diffusive random lasing, typically pulse widths of a few nanoseconds can be employed. A combination of varying opacity neutral density filters was placed as attenuators, in order to tune the excitation energy E_p. A glass window after the attenuators in the laser

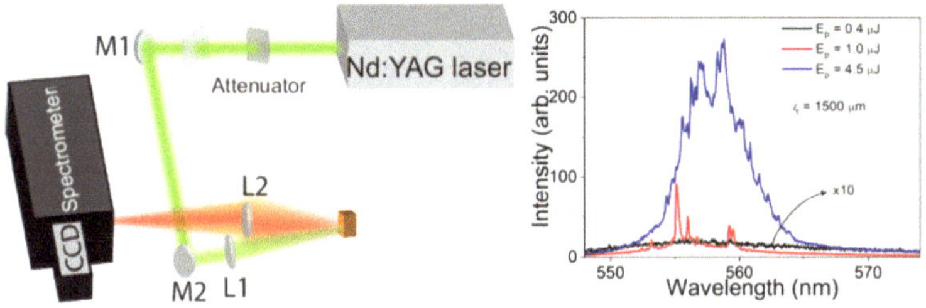

Figure 4.1 Schematic of a typical experimental setup for studying random lasing emission. The beam from an Nd:YAG laser is attenuated for controlling excitation energy E_p, and then partially reflected off a window to record pump energy on a photodiode (not shown here). $M1$ and $M2$ are mirrors that steer the beam onto lens $L1$ which focuses the pump beam on the sample. The lens $L2$ collects the emission for spectral analysis using a monochromator-CCD device. The right panel illustrates the emission spectra at the mentioned excitation energies.

beam reflected a fraction of the beam onto a photodetector (not shown here) to record the pulse-to-pulse fluctuations in E_p. The excitation laser beam is focused onto the face of the cuvette using the focusing lens $L1$ (focal length = 200 mm), placed on a translation stage. The random laser emission was collected using the lens $L2$ (focal length = 75 mm) and imaged onto the input slit of a monochromator (focal length = 50 cm). The monochromator (Czerny–Turner configuration) comprised a 1200 lines/mm diffraction grating that helped in spectrally resolving the emission, which was made incident on an intensified-CCD pre-configured to the monochromator. Under these measurement conditions, a spectral bandwidth of ~26 nm could be collected on the CCD with a fine resolution of ~0.12 nm. The plot exhibits three spectra at the three mentioned excitation energies. At sub-threshold excitation (black line, emphasized by a factor of 10), the emission is broadband and low intensity. Just above the lasing threshold, at 1 μJ (red line), the spectrum shows ultranarrow coherent emission modes at random wavelengths. At still higher pump energy (blue line), the spectrum grows in intensity, but loses its ultranarrow character. This behavior is traced to the self-averaging that sets in at higher energies (Uppu and Mujumdar, 2011).

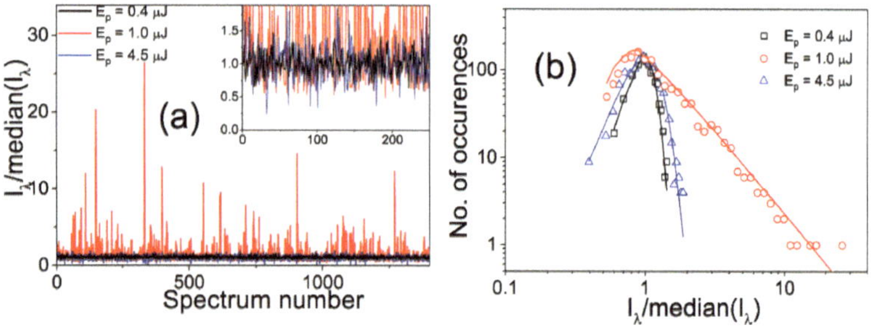

Figure 4.2 (a) It depicts the series of measured intensity values at $\lambda = 557$ nm, from a random laser with $\ell^* \sim 1500$ μm and pump energy $E_p \sim 0.4$ μJ (black curve), $E_p \sim 1.0$ μJ (red curve), and $E_p \sim 4.5$ μJ (blue curve). The inset shows an emphasized view of the low-intensity data, revealing the weak fluctuations at $E_p \sim 0.4$ μJ and 4.5 μJ. (b) It shows thc distribution of emission intensities on a log-log scale at the three excitation energies. The solid lines represent the fits to the $\alpha-$stable distribution.

4.1.2 Statistical Data

To record the statistics, we measure the pulse-to-pulse fluctuations in the emission intensity at the emission maximum, i.e., $\lambda \sim 557$ nm. Figure 4.2(a) shows the experimentally measured series of the intensities over 2500 spectra at three different excitation energies: $E_p \sim 0.4$ μJ (black curve), $E_p \sim 1.0$ μJ (red curve), and $E_p \sim 4.5$ μJ (blue curve). In order to enable comparison across different E_p, the intensities I_λ are normalized to the median of the fluctuations. Under weak excitation ($E_p \sim 0.4$ μJ, black curve), where the random laser is still sub-threshold, only weak fluctuations from the median are noticed. These weak fluctuations arising from the fluorescence are better seen in the inset, where the low-intensity region is magnified. At $E_p \sim 1.0$ μJ (red curve), strong unbounded fluctuations appear as a consequence of the occurrence of coherent peaks in the emission. Upon further increase of E_p (blue curve), a self-averaging process sets in (Uppu and Mujumdar, 2011) that inhibit these strong fluctuations. Nonetheless, these fluctuations are still mildly stronger than those observed in fluorescence, seen in the inset.

4.2 Lévy Fluctuations in Random Lasers

In the above series of intensity, preliminary statistical analysis immediately exhibits the divergence of the second moment or the variance of the distribution, which is a signature of power-law fluctuations or Lévy fluctuations. The origin of Lévy fluctuations in this system is easy to understand. In the process of diffusion-mediated amplification, the spontaneous emission photons undergo the Brownian random walk before exiting the system (Berger et al., 1997; Mujumdar et al., 2004), and the path length L distribution of the photons is given by (Redner, 2001) $P(L) = \frac{1}{\langle L\rangle}e^{-L/\langle L\rangle}$, where $\langle L\rangle$ is called the mean first-passage length of the exiting photons. The process of amplification in the medium scales exponentially with L, yielding the emission intensity $I(L) = I_0 e^{L/\ell_g(\lambda)}$, where I_0 is the initial intensity of the random walker (= 1, a single spontaneous emission photon) and $\ell_g(\lambda)$ is the gain length of the medium at the wavelength λ. The ℓ_g is the mean distance over which a photon amplifies to 2 and is determined as: $\ell_g = \frac{1}{\sigma_e(\lambda)N_1 - \sigma_a(\lambda)N_0}$, where $\sigma_e(\lambda)$ and $\sigma_a(\lambda)$ are the emission and absorption cross-sections of the gain medium, respectively. N_1 and N_0 are the population density of the molecules in the excited and ground states, respectively.

The cumulative distribution function of the intensity $F(I)$ is obtained through a change of variable process as follows:

$$F(I) = \int_{\ell_g \log(I/I_0)}^{\infty} \frac{1}{\langle L\rangle} e^{-L/\langle L\rangle} dL$$

$$F(I) = \left(\frac{I_0}{I}\right)^{\mu}; \qquad \mu \equiv \ell_g./\langle L\rangle \tag{4.1}$$

Thus, the probability distribution turns out to be

$$P(I) = \mu \frac{{I_0}^{\mu}}{I^{\mu+1}}. \tag{4.2}$$

Accordingly, the intensity distribution follows a power-law (Lévy) distribution with the exponent μ (Sharma et al., 2006a,b; Lepri et al., 2007), which depends on the disorder strength ℓ^*, excitation energy E_p and the excitation volume in the sample. Power-law distributions support infinite variance when $\mu < 2$ (Lévy,

1937), which manifests unbounded fluctuations in the variables, as mentioned earlier.

Although the power-law distribution originates from the exponential amplification of the spontaneous emission photons, it is not possible to measure the intensities of individual photons. In reality, the experiment measures the sum of intensities of the spontaneous emission photons in a small spectral bandwidth, determined by the spectral resolution. Therefore, the measured intensity is actually the sum of many photons, each of which is a Lévy variable. Such sums of Lévy variables are called Lévy sums, the distribution of which is given by the α-stable distribution (Gnedenko and Kolmogorov, 1954). It models a heavy-tailed distribution with an asymptotic tail exponent α, whose value indicates the statistical nature of the distribution of the Lévy sums, with $0 < \alpha \leq 2$. For $\alpha = 2$, the statistics are Gaussian in nature, while for $\alpha < 2$, the distributions are Lévy. Smaller the α, stronger the Lévyness and hence, larger the fluctuations.

4.2.1 The α-Stable Distribution

The α-stable distribution is completely described by a set of four parameters, namely, the Tail parameter $\alpha \in (0, 2]$, the skewness parameter $\beta \in [-1, 1]$, the Scale parameter $\gamma \in (0, \infty)$, and the Location parameter $\delta \in (-\infty, \infty)$. Barring a few special cases, the distribution lacks a closed-form formula and is generally represented through its characteristic function $\phi(t)$, which is the inverse Fourier transform of the probability density function:

$$\phi(t) = \int_{R} e^{itx} P(x)\boldsymbol{d}x. \tag{4.3}$$

Multiple parametrizations have been used for the α-stable law, out of which we employed one called $S(\alpha, \beta, \gamma, \delta; 0)$, denoted as: (Nolan, 2013)

$$\ln \phi(t) = \begin{cases} -\gamma^{\alpha}|t|^{\alpha}\left[1 + i\beta\left(\tan\frac{\pi\alpha}{2}\right)\mathrm{sign}(t)(|\gamma t|^{1-\alpha} - 1)\right] + i\delta t & \alpha \neq 1 \\ -\gamma|t|\left[1 + i\beta\frac{2}{\pi}\mathrm{sign}(t)\log(\gamma|t|)\right] + i\delta t & \alpha = 1 \end{cases}. \tag{4.4}$$

In our analysis, toward the parameter estimation, we used a regression-based algorithm (Koutrouvelis, 1980), which is abundantly employed in econophysics and financial statistics. The algorithm is known to be fast and robust.

Figure 4.2(b) shows the intensity distributions on a log-log scale at the three pump energies, along with the corresponding $\alpha-$stable fits. At $E_p = 0.4$ μJ (open black squares), the fit yields $\alpha = 2.0$, endorsing the Gaussian regime of fluctuations in the sub-threshold emission, which is as expected for fluorescence. At $E_p = 1.0$ μJ (open red circles), the occurrence of distinct coherent peaks manifests as strong Lévy fluctuations with $\alpha \sim 0.9$. At a higher E_p of 4.5 μJ, the intensity fluctuations are reduced, yielding $\alpha \sim 1.95$, suggesting near-Gaussian statistics.

Figure 4.3(a) illustrates the systematic dependence of the power-law exponent α (empty black squares) on the excitation energy E_p. Initially, a rapid fall from $\alpha = 2$ to $\alpha \sim 0.9$ is seen to occur at the random lasing threshold E_{th}. Thereafter, upon increasing E_p above the threshold, the self-averaging process subdues fluctuations, which manifest as a gradual increase in α. This realizes a slow crossover of the statistical nature from the strong Lévy regime to the second Gaussian regime (Uppu et al., 2012), with the minimum corresponding to the E_{th}.

4.2.2 α as a Threshold Identifier

The above behavior of α suggests an interesting utility in determining the random lasing threshold, and thus the criticality. In the field, different definitions for the random lasing threshold in diffusive and coherent random lasers have been used to suit the different situations of investigation. In the case of a coherent random laser, a measure of the propensity to emit light in the ultranarrow peaks, called the *probability of random lasing* (P_{RL}), has been defined (Fallert et al., 2009; Tiwari et al., 2012). It is determined by the ratio of the number of spectra demonstrating lasing modes to the total number of collected spectra, and the E_p at which the probability becomes non-zero identifies the E_{th}. In the diffusive random laser, the threshold is typically identified using either the input-output curve or the bandwidth collapse (Lawandy et al., 1994). In

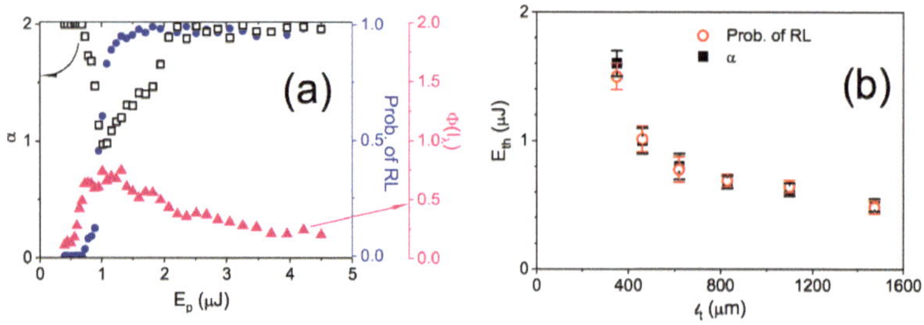

Figure 4.3 (a) It shows the variation of the probability of random lasing P_{RL} (blue circles) and fluctuations of intensity (magenta triangles) with E_p. The α variation is shown by the black squares. (b) It shows a comparison of the lasing threshold E_{th} estimated from the α−statistics (black squares) and from the measurement of the probability of random lasing P_{RL} (red circles).

either case of coherent or diffusive lasing, the onset of random lasing occurs concurrently with a pronounced increase in intensity fluctuations (van der Molen et al., 2006; Uppu and Mujumdar, 2010; Zhu et al., 2012). This rise in fluctuations corresponds to the Gaussian-Lévy transition, which can be measured through α. Figure 4.3(a) illustrates the dependence of the (P_{RL}) on E_p (blue circles). P_{RL} increases to a non-zero value around 0.5μJ indicating the onset of lasing. This increase in the occurrence of ultranarrow peaks results in stronger fluctuations resulting in the drop in α below 2, which also identifies the E_{th}. Figure 4.3(b) shows a comparison between the estimated E_{th} from P_{RL} (red circles) and the Gaussian-Lévy transition (black squares), where an excellent agreement can be seen. Interestingly, the E_{th} for the stronger disorder could not be estimated using (P_{RL}), because of the loss in contrast to the ultranarrow peaks, which makes them indiscernible. However, the α-exponent could still estimate the E_{th} even when coherent peaks are absent. Similarly, in another set of experiments, the threshold energies for diffusive random lasing systems were also accurately estimated by the α-exponent. Clearly, this measure of the statistical nature of the fluctuations is not only a good indicator for lasing threshold, but a robust and universal one too.

4.3 Estimation of the Gaussian-Lévy Transition

From the description presented in the earlier section, α is the tail parameter of the distribution of Lévy sum of intensities S_N, where $S_N = \sum_{i=1}^{N} I_i$, such that $P(I_i) = \mu \frac{I_0{}^{\mu}}{I_i^{\mu+1}}$. Here, I_i is the amplified intensity of the N individual emission photons and $P(I_i)$ is their probability distribution. Hence, the estimation of α requires the knowledge of μ and N. In systems with $\mu < 2$, the values of μ and α turn out to be equal. Thus, in order to identify the Gaussian-Lévy transition, it suffices to estimate the pump energy at which μ falls below 2.

The estimation of $\mu \equiv \ell_g / \langle L \rangle$ requires the calculation of ℓ_g and the mean first-passage length $\langle L \rangle$, or equivalently the mean first-passage time $\langle \tau \rangle = \langle L \rangle / v$, where v is the speed of light in the medium. This is theoretically accessible by solving the diffusion equation over a relevant geometry (Redner, 2001). For typical random lasing experiments, this geometry is best approximated by a cylinder with walls that are perfectly absorbing. Figure 4.4 schematizes the photon diffusion out of a cylindrical volume of length L and radius a. The photons can exit from either the front face $z = 0$, the back face $z = L$, or the surface $r = a$. Experimentally, however, the detector only collects photons exiting the front face, i.e., $z = 0$. The calculation of $\langle L(\vec{r}') \rangle$ in cylindrical coordinates with $\vec{r}' = (\rho', z', \phi')$ is detailed in Reference (Uppu and Mujumdar, 2013). The final expression is

$$\begin{aligned}\langle L(\vec{r}') \rangle = & \frac{v}{\pi_0(\vec{r}')\beta} \sum_{n=1}^{\infty} \frac{(x_{0n}\beta)}{x_{0n}^2 J_1(x_{0n})} J_0(x_{0n}\beta\rho') \\ & \times [\coth(x_{0n}\beta) \sinh(x_{0n}\beta(1 - z')) \\ & -(1 - z') \cosh(x_{0n}\beta(1 - z'))]. \end{aligned} \tag{4.5}$$

Here, x_{0n} is the nth zero of the zeroth-order Bessel function of the first kind, J_0. The parameter β is the aspect ratio of the cylinder $(= L/a)$ and $\pi_0(\vec{r}')$ represents the splitting probability for $z = 0$, i.e., the probability for photons to exit out of the front face, and is given by

$$\pi_0(\vec{r}') = 2 \sum_{n=1}^{\infty} \frac{(x_{0n}\beta)}{x_{0n} J_1(x_{0n})} J_0(x_{0n}\beta\rho') \sinh[x_{0n}\beta(1 - z')]. \tag{4.6}$$

From Eqs. 4.5 and 4.6, $\langle L \rangle$ can be calculated, given β. The β quantifies the shape of the excitation subvolume, which is determined by ℓ^*, E_p, the focal spot diameter of the pump, and the concentration of the dye medium. Since β is analytically intractable, we employ a numerical simulation for the photon transport using Monte Carlo techniques (Mujumdar and Ramachandran, 2000; Mujumdar et al., 2004; Uppu and Mujumdar, 2013).

4.3.1 Photon Transport Monte Carlo Simulations

The simulation volume is a 1 cm × 1 cm × 1 cm cube discretized into 2 μm × 2 μm × 2 μm cells. The volume is filled with a 2.5 mM Rhodamine 6G solution with an absorption cross-section σ_a. The disorder is realized by assuming that this amplifying medium contains a random distribution of point scatterers with number density N_s and scattering cross-section σ_s, giving the scattering mean free path $\ell_s \equiv 1/N_s\sigma_s$. The excitation pulse is realized by simulating bunches of pump photons ($\lambda = 532$ nm) launched perpendicularly (see Fig. 4.4) from the center of the front face with initial angular coordinates $(\theta_0, \phi_0) = (0{,}0)$. The starting positions of the photon bunches at the front face are Gaussian-distributed with an FWHM of 60 μm. Each bunch comprises N_{pb} number of photons at launch, which undergoes a three-dimensional random walk within the simulation space. The step length l of the random walk is exponentially distributed as $P(l) = (1/\ell_s)\exp(-l/\ell_s)$. The scattering event at the end of each rectilinear step is simulated by choosing another direction (θ_i, ϕ_i) that is uniformly distributed over 4π. This uniform distribution assumes isotropic scattering, but anisotropy can be incorporated through the Henyey–Greenstein or Mie phase functions (Henyey and Greenstein, 1941; Irvine, 1965; Toublanc, 1996).

During the propagation, a fraction of the N_{pb} gets absorbed, depending on the absorption length of photons $\ell_{abs}(i, j, k) = [N_0(i, j, k)\sigma_{abs}(\lambda_p = 532 \text{ nm})]^{-1}$, where $N_0(i, j, k)$ is the number of ground-state molecules in the $(i, j, k)^{th}$ cell. A proportional number of ground-state molecules are added to the excited state $N_1(i, j, k)$. This propagation and absorption process is repeated until either $N_{pb} < 1$ or the bunch exits the system. The number of bunches

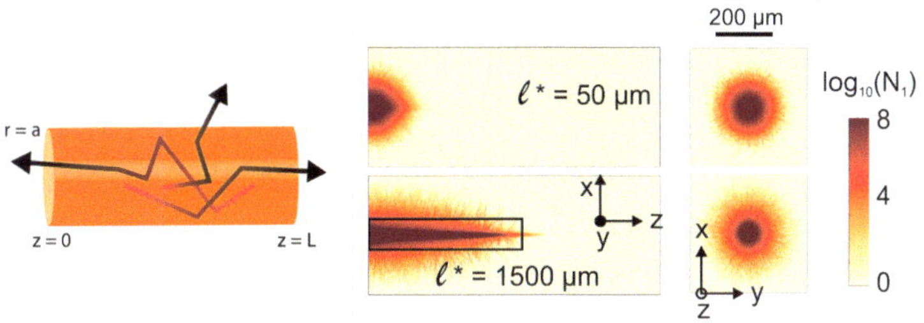

Figure 4.4 The left figure schematizes photons diffusing out of a cylinder of length L and radius a. All faces of the cylinder are assumed to be perfectly absorbing, so photons can only exit out of the walls. The right image depicts the excited state population N_1 (on a log scale) via the longitudinal and transverse cross-sections at two disorder strengths. The black rectangle in the $\ell^* = 1500$ μm system represents the smallest perfectly absorbing cylinder.

decides the desired pump energy E_p. The spatial distribution of $N_1(i, j, k)$ replicates the pump volume under the excitation of a short laser pulse.

Figure 4.4 depicts the longitudinal and transverse cross-sections of the excited state population (N_1) at $E_p = 4.5$ μJ in systems with $\ell^* = 50$ μm and $\ell^* = 1500$ μm. The excitation is nearly axisymmetric and can be approximated by a cylinder. The black outlined rectangle shown in the transverse profile of the $\ell^* = 1500$ μm system represents the cylinder in this case. The β of this cylinder thus determines $\langle L \rangle$.

4.3.2 The First Gaussian–Lévy Transition.

Figure 4.5(a) depicts the behavior of the splitting probability π_0 as a function of the starting positions in systems with different aspect ratios: $\beta = 4$ (black curve), $\beta = 8$ (dashed red curve), and $\beta = 12$ (dotted blue curve). The exit probability for photons generated deep inside the cylinder decreases with increasing β. This is due to the higher absorption probability at the walls of a long cylinder during the random walk. Figure 4.5(b) illustrates the mean first-passage time $\tau_0 = \langle L \rangle / v$ as a function of the starting positions. Clearly, systems with larger β support longer times in the pump volume,

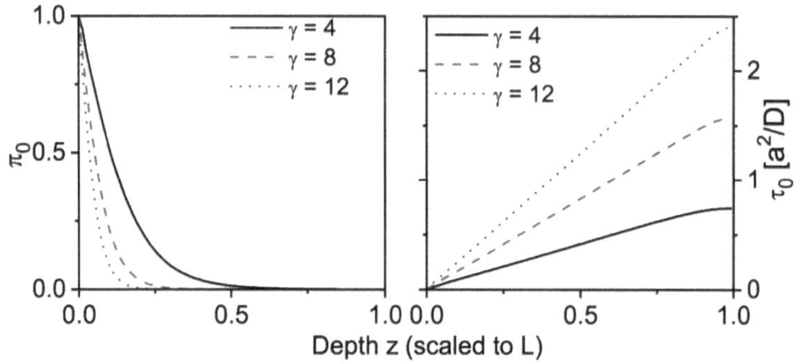

Figure 4.5 The left panel shows the splitting probability at the front face π_0 for three cylinders with aspect ratios: $\beta = 4$ (black curve), $\beta = 8$ (dashed red curve), and $\beta = 12$ (dotted blue curve). The right panel shows the mean first-passage time τ_0 (scaled by a^2/D, where D is the diffusion coefficient) of photons diffusing through the front face for the same β.

or equivalently, longer paths. Thus, these calculations indicate that systems with large β support rare but long photon paths that exit from the front face. In the presence of inversion, such paths amplify strongly and result in strong ultranarrow peaks in emission. As evident from Fig. 4.4, such large β occurs in the case of weak scattering systems, which indeed are the systems that demonstrate strong ultranarrow peaks in emission (see Fig. 4.1(b)).

The β then allows for the calculation of $\langle L \rangle$. The photon transport simulations also yield a mean-field estimate of ℓ_g from the mean excited-state and ground-state populations over the cylinder. Thus, the power-law index $\mu = \ell_g/\langle L \rangle$ is calculated for various combinations of ℓ_s and E_p. Figure 4.6(a) depicts the variation of the μ with E_p in systems with different disorder strengths. The horizontal dotted red line at $\mu = 2$ demarcates the Lévy regime below the line from the Gaussian regime above. Increasing E_p results in a monotonic decrease of μ at all disorder strengths. The intersection of the $\mu = 2$ line with the monotonically decreasing profile provides the threshold pump energy at which the system crosses from the Gaussian–Lévy domain. Figure 4.6(b) reveals a comparison of the threshold energy estimates (black squares) with the experimentally measured E_{th} (red circles). The E_{th} decreases rapidly at shorter ℓ^* and becomes almost independent of the

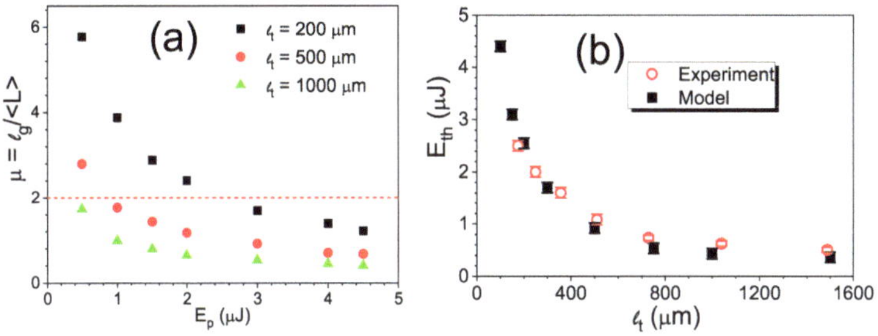

Figure 4.6 (a) It depicts the computed Lévy exponents μ as a function of E_p at the three listed disorder strengths. The horizontal dashed red line at $\mu = 2$ separates the lower Lévy domain from the upper Gaussian domain ($\mu > 2$). (b) Comparison of numerically estimated (black squares) random lasing thresholds E_{th} with the experimentally measured (red circles) threshold, showing an excellent agreement.

disorder strength in the weaker samples. A very good quantitative agreement is seen between the experimental and the calculated threshold energies across the entire range of disorder strengths (Uppu and Mujumdar, 2013).

As far as the first Gaussian–Lévy transition is concerned, the agreement between experiments and theory is excellent. However, it can be immediately seen that with increasing E_p, the profile of μ will enter deeper into the Lévy regime. Contrarily, the experiments show a recovery curve, marking a Lévy-Gaussian crossover at high pump energies. Clearly, this departure requires further insights and is clarified in the next section.

4.4 Exponentially Tempered Lévy Sums

The crux of the departure from experiments lies in the unrealistic unboundedness of the second moment of the Lévy process, which implies that infinite step size is permissible in the process. But such infinities are not tolerated by any real physical system. Accordingly, in random lasers, a finite bound on the intensity is clamped by two factors, namely, the finite gain available for amplification, and the finite system size, which can support only a finitely long path for the diffusing photon. Clearly, the resulting amplification cannot sustain infinite fluctuations in intensity in the random laser.

4.4.1 Truncated Lévy Flights

In order to resolve this 'infinity paradox' in applications of Lévy processes to real systems, two models have been proposed: (*a*) Lévy walks (Shlesinger et al., 1993) and (*b*) Truncated Lévy Flights (TLF) (Mantegna and Stanley, 1994). The former penalizes extra-long displacements and is applicable in certain processes. In our case of the random laser, the latter model is more relevant. The TLF resolves the infinity by forcibly truncating the power-law distribution above a cut-off intensity I_{tr}, leading to a probability distribution as:

$$P(I) = \frac{\mu}{\left[1 - \left(\frac{I_0}{I_{tr}}\right)^{\mu}\right]} \frac{I_0^{\mu}}{I^{\mu+1}}; \qquad I_0 \leq I \leq I_{tr}. \tag{4.7}$$

The imposed cut-off limits the divergence in the sample variance indicating a departure from the unbound fluctuations. The introduction of finite variance, then, affects the tail exponent α of the sums of the Lévy variables (Mantegna and Stanley, 1994, 2000). For a given μ, the extent of the effect is determined by I_{tr} and the number of summands N_m. Figure 4.7(a) shows the dependence of α on the number of summands N_m at different values of I_{tr}: (a) Untruncated, i.e., $I_{tr} \rightarrow \infty$ (black squares), (b) $I_{tr} = 10^6$ (blue circles), and (c) $I_{tr} = 10^5$ (red triangles). The calculation of α was performed as follows. A number N_m of power-law distributed random variables with $\mu = 1.2$ with a given I_{tr} were summed to generate one sum. 2000 such sums were computed and the tail exponent α of their distribution was estimated. In the untruncated case, $\alpha \sim \mu$ for all values of N_m, as expected (Samarodnitsky and Taqqu, 1994). A weak truncation ($I_{tr} = 10^6$) results in a deviation from $\alpha = 1.2$ at $N_m \sim 10^5$, beyond which, α increases and eventually converges to the Gaussian regime ($\alpha = 2$) at $N_m > 10^7$. At smaller cut-off values ($I_{tr} = 10^5$), this convergence is faster. This crossover to a Gaussian regime due to an imposed truncation has been extensively studied in various fields (Mantegna and Stanley, 1994; Burroughs and Tebbens, 2001; Bruno et al., 2004; del Castillo-Negrete, 2009; Sibatov and Uchaikin, 2011; Matacz, 2000). In the random laser, N_m corresponds

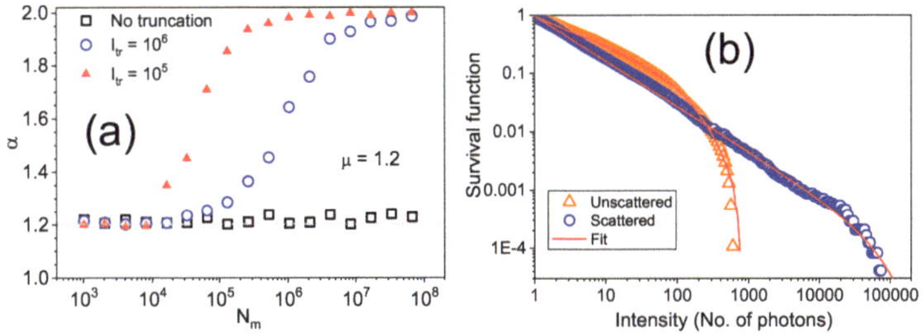

Figure 4.7 (a) It illustrates the effect of finite truncation on the Lévy nature of the sums of power-law variables, quantified using α. The harder the truncation, the faster the convergence to a Gaussian. (b) The plot shows the survival functions of emission intensities for ballistic (orange triangles) and multiplies scattered (blue circles) photons in a system with $\ell^* =$ 1500 μm. The red curves are the fits to an exponentially tempered power-law function.

to the number of spontaneous emission photons that comprise the intensity at a wavelength λ.

Although the introduction of a cut-off on the power-law distribution resolves the infinities, it creates a singularity in the derivative of the probability density function at I_{tr}. Physical systems require all functions to be smooth and cannot support such singularities. This issue is resolved by a smooth cut-off, wherein the probability of large jumps is exponentially tapered, so that all moments exist (Rosinski, 2007). This process has been applied successfully in various studies (Meerschaert et al., 2008; Stanislavsky et al., 2008; Gajda and Magdziarz, 2010; Chakrabarty and Meerschaert, 2011; Oshanin et al., 2011). Here, we use these exponentially tempered Lévy sums (ETLS) to model the random laser. The probability density function $P(I)$ and the survival function $\overline{F}(I)$ of the exponentially tempered power law are written as:

$$P(I) = \gamma I^{-(\mu+1)} e^{-\beta I}[\mu + I\beta]; \qquad I \geq I_0 \tag{4.8}$$

$$\overline{F}(I) = \gamma I^{-\mu} e^{-\beta I}; \qquad I \geq I_0 \tag{4.9}$$

where μ is the power-law exponent, γ is the scale parameter (= 1 in our analysis involving normalized distributions), and β controls the exponential tempering. The tempering parameter can be used to define the cut-off intensity $I_{tr} \equiv 1/\beta$.

4.4.2 Parameter Estimation for Lévy Sums

In order to calculate the parameters μ, N_m, and I_{tr} $(= 1/\beta)$, we perform PTMC simulations, discussed in section 4.3.1, to study the characteristics of emission photons. Figure 4.7(b) depicts the survival functions $\overline{F}(I)$ on a log-log scale of the ballistic (unscattered) photons (orange triangles) and the multiply scattered photon (blue circles) in the system with $\ell^* = 1500$ μm. The distributions trail off to zero after an initial linear decay. The intensity at which the decay sets in is lower for the unscattered photons. A larger linear region is seen in the distribution of the multiply scattered photons. The red curves depict the fit to the exponentially tempered power-laws using a Maximum Likelihood method (Meerschaert et al., 2012). The fits yield μ of 0.52 and 0.7 and $I_{tr} = \beta^{-1}$ of 700 and 4×10^6 for the unscattered and scattered photons respectively. The unscattered photons experience a very strong cut-off, and hence the fluctuations in the Lévy sums depend primarily on the multiply scattered photons.

Apart from μ and β, the estimate for the N_m is also provided by the PTMC simulations, wherein an estimate for the total number of spontaneous emission events required to completely de-excite the system can be found. The fraction of these events at the emission wavelength $\lambda = 557$ nm is an estimate for N_m for our system.

Figure 4.8 illustrates the dependence of the estimated power-law parameters on the disorder strength. Figure 4.8(a) depicts the variation of μ (open blue symbols; right Y-axis) and $1/\beta(= I_{tr})$ (filled black symbols; left Y-axis) with E_p. Two systems with different scattering strengths, namely, $\ell^* = 350$ μm (triangle markers) and $\ell^* = 1500$ μm (circle markers) are illustrated. A monotonic decrease in μ with increasing E_p is seen. Furthermore, weaker scattering systems possess smaller magnitudes of μ. A fast increase followed by a strong limitation of I_{tr} is seen with increasing E_p. Furthermore, I_{tr} is consistently smaller for the stronger scattering system. This is a consequence of the weaker local inversion caused by the uniform redistribution of pump energy over the pump volume. Figure 4.8(b) depicts the fraction of scattered spontaneous emission events f_m among the total modes N_m, i.e., $f_m N_m$, at the two disorder strengths with the corresponding

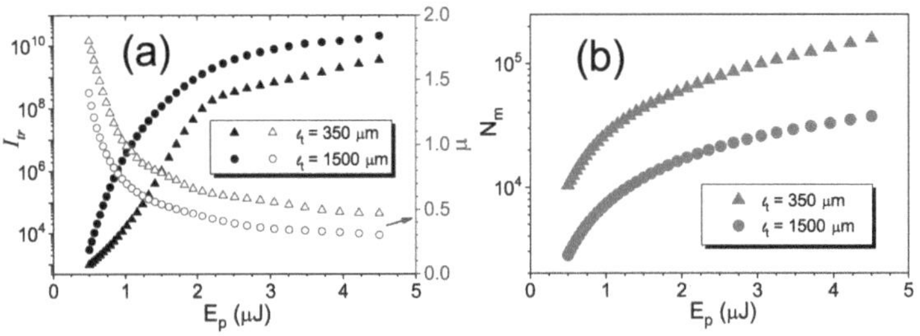

Figure 4.8 (a) It traces the dependence of μ (open blue symbols) and I_{tr} (filled black symbols) of the multiply scattered photons on E_p. Two different disorder strengths are studied here: $\ell^* = 350$ μm (triangles) and $\ell^* = 1500$ μm (circles). (b) It depicts the fraction of scattered photons in N_m with E_p in these two systems.

markers. A larger number of photons are required to de-excite the strong scattering system, owing to the higher exit probability coupled with the weaker amplification.

Finally, for the given μ, N_m, and I_{tr}, the ETLS can be generated using the rejection sampling technique. A power-law distributed random variable $y = u^{-1/\mu}$ is generated, where $u \in (0, 1]$ is a uniformly distributed random variable. This y is accepted as an exponentially tempered power-law variable if, for another uniform random variable $v \in (0, 1]$, $f(y)/g(y) \leq v$. Here, $f(y) = x^{-\mu-1}e^{-\beta x}(\mu+x\beta)$ and $g(y) = \mu/(x^{\mu+1})$. A total of N_m such variables are generated and summed, and 2000 such sums were fit to an α-stable distribution to yield the tail exponent α.

Figure 4.9 compares the experimentally measured statistics (left panel) and numerically estimated α (right panel) across different values of ℓ^* and E_p. The $\alpha = 2$ region, i.e., Gaussian statistics are indicated by the white portion of the colormap. Darker shades on the color bar indicate smaller α and hence stronger Lévy nature of fluctuations. Three α-maps are presented for three different focal spot sizes, namely, 60 μm (top subpanel), 120 μm (middle subpanel), and 240 μm (bottom subpanel). An excellent agreement is evident between the measured and computed statistics in intensity fluctuations across the entire parameter space. This indicates that the ETLS model captures the random laser dynamics accurately. For the smallest spot size, the E_p over which the Lévy

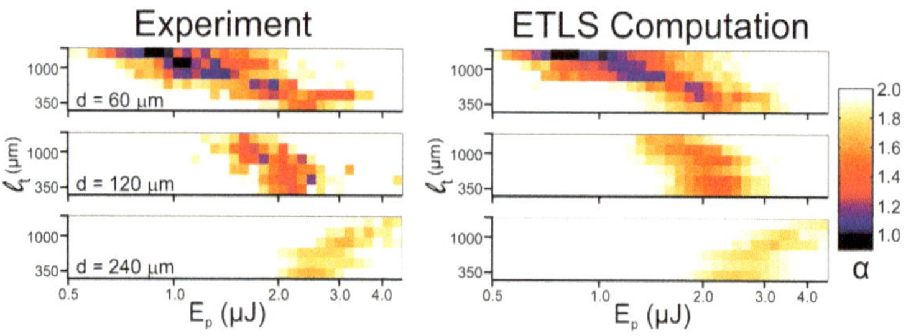

Figure 4.9 The left panel corresponds to the experimentally measured α over a range of values of E_p and ℓ^*. The right panel demonstrates the calculated α using the exponentially tempered Lévy sum approach. The top, middle, and bottom panels depict the α-maps from a system with pump focal diameters $d_p \sim 60$ μm, 120 μm, and 240 μm, respectively.

fluctuations occur reduces with disorder strength. Moreover, the range of E_p over which Lévy fluctuations occur increases with weakening disorder strength. For the moderate spot size, this range does not vary strongly. For the largest spot size (240 μm), the trends are reversed. The Lévy fluctuations occur at stronger disorder at lower E_p. In weaker disorders, the higher pump is required to trigger Lévy fluctuations. Also, the Lévyness in the fluctuations is weaker at a larger spot size, due to the self-averaging effects mentioned earlier.

4.5 Extreme Events in Random Lasers

The Lévy sum S_N is constituted of the N intensities of spontaneous emission events that undergo amplification, i.e., $S_N = \sum_k = 1^N I_k$, where I_k is the intensity of the k-th event. The limit distribution of S_N when $N \to \infty$ depends on the value of the Lévy exponent μ. As the underlying Lévy distribution results in the divergence of the mean and/or the variance of the distribution for $\mu < 2$, the central limit theorem is not applicable. Generalized central limit theorem (GCLT) can be employed to capture the scaling of S_N as $N \to \infty$ (Gnedenko and Kolmogorov, 1954). The GCLT predicts separate scaling laws for the two scenarios: (a) $\mu < 1$, where both the mean and the variance diverge, and (b) $1 \leq \mu < 2$, where only the variance diverges. The predicted scaling using the GCLT is

$$S_N = I_0 N^{1/\mu} Z; \qquad 0 < \mu < 1 \tag{4.10}$$

$$S_N = mN + I_0 N^{1/\mu} Z; \qquad 1 \leq \mu < 2, \tag{4.11}$$

where m is the mean of the distribution, Z is a random variable from the $\alpha-$stable distribution with $\alpha = \mu$ for an untruncated distribution, and I_0 is the lower limit of the power-law distribution $P(I) = \mu I_0^{\mu}/I^{\mu+1}$. These scaling laws indicate the possibility of large unbounded values of sum even as $N \to \infty$. In contrast, if $\mu > 2$, the scaling law for S_N is

$$S_N = mN + \sigma\sqrt{(N)}Z \tag{4.12}$$

where σ^2 is the variance. Therefore, for $\mu > 2$, the sum S_N scales as $\sqrt{N}$ as predicted from the central limit theorem.

An interesting question to pose is: How many summands contribute to the sum that causes the unbounded behavior? We can trace the contribution of the individual summands by studying the scaling of the order statistics. Given the set of summand intensities I_k, the $n-$th order statistic is the $n-$th largest value in the set and is represented as $I^{(n)}$. By analyzing the order statistics, we can derive that the $n-$th order statistic scales as

$$I^{(n)} = I_0 \left(\frac{\mu N + 1}{\mu n + 1}\right)^{1/\mu}. \tag{4.13}$$

Interestingly, the scaling of the largest summand, i.e., the first-order statistic $I^{(1)} \approx I_0 N^{1/\mu}$. This scaling is the same as that of the sum S_N when $\mu < 1$, which establishes that a single summand can be as large as the sum itself. Such rare extreme summands are the dominant factor in the Lévy regime of random lasers.

The random laser system with $\ell^* = 1500$ μm excited ~ 1 μJ is close to the random laser threshold (see Fig. 4.1). The computed power-law tail exponent of the Lévy variables in this system was found to be $\mu = 0.7$ (see Fig. 4.9). Furthermore, the I_{tr} was large so there was a chance of the Gaussian crossover. This small value of μ along with the large I_{tr} implies that large fluctuations in the individual terms can occur. And these fluctuations can be of the same order as the Lévy sum itself.

Figure 4.10(a) illustrates a series of computed 2500 realizations of Lévy sums from the system with the said parameters. The

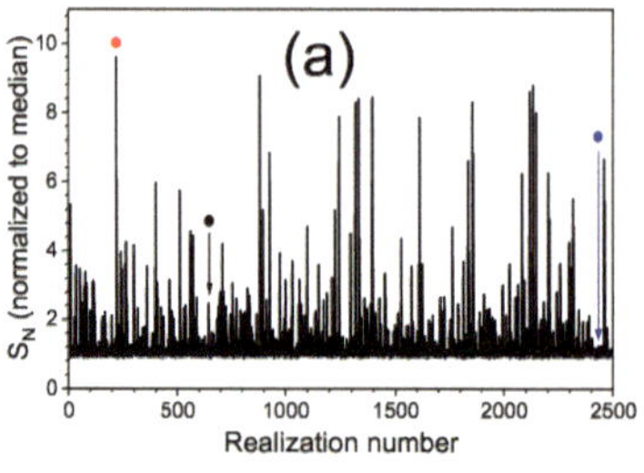

Figure 4.10 (a) It depicts a series of 2500 executions of Lévy sums realized for a random laser with $\ell^* = 1500$ μm, at an excitation energy of $E_p = 1$ μJ. Red, blue, and black markers indicate three particular realizations with varying levels of deviations from the median. (b) It depicts the normalized cumulative sum of the independent Lévy variables that participate in the sums marked in figure (a) in like colors.

simulations provide access to the individual terms that comprise the Lévy sum. Three realizations are marked in the figure with red, black, and blue circles, which are representative of large, moderate, and small fluctuations in the system. Figure 4.10(b) shows the cumulative sum of the individual terms for these three realizations. The cumulative sums are normalized to the value of the sum for comparison. For the realizations marked with red and black markers, single large jumps are evident in the plot. These are the realizations where strong deviations from the median occur. These large jumps arise from a single random variable whose value is comparable to the sum. The ratio of the largest summand to the sum, i.e., $I^{(1)}/S_N$ is $\sim$ 0.9, 0.6, and 0.06 for the realizations marked with red, black, and blue markers. Therefore, it is evident that large fluctuations in sums imply a large contribution from a single rare large term.

The relevant experimental data are illustrated in Fig. 4.11(a). The plot depicts these measured intensity fluctuations at $\lambda = 557$ nm from the system with the same ℓ^* and E_p as the simulations. Large fluctuations in emission intensity are seen. Three intensity values with varying degrees of fluctuations from the median are marked out with red, black, and blue markers in the figure. Figure 4.11(b) shows the experimentally measured spectra corresponding to these three intensity values in like colors. The spectra are offset

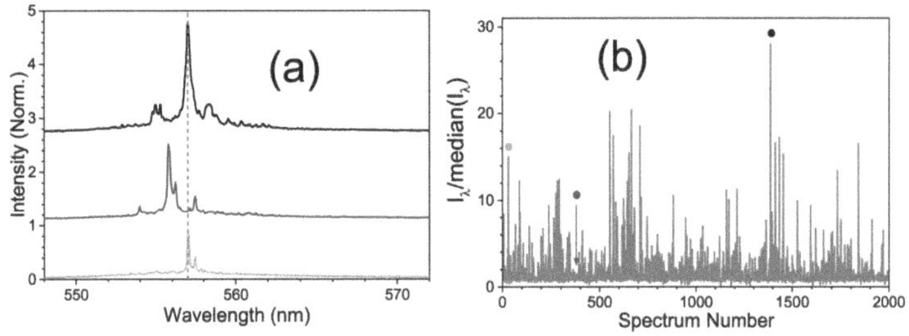

Figure 4.11 (a) It depicts three captured spectra, which exhibited ultranarrow peaks, from a system with ℓ^* ~1500 μm when excited at E_p ~ 1 μJ. (b) It illustrates the measured intensity fluctuations at $\lambda =$ 557 nm in this system. The red, blue, and black circles mark the particular measurements, which correspond to the spectra in like colors.

vertically for clarity, and the vertical black dotted line marks $\lambda =$ 557 nm. Strong ultranarrow peaks at this wavelength are visible in red and black curves. Ultranarrow peaks do occur in the blue curve too, however, at a wavelength other than $\lambda = 557$ nm. This directly correlates the strong intensity fluctuations at a wavelength with the occurrence of coherent peaks at that particular wavelength. From Fig. 4.10 and Fig. 4.11, we can conclude that the coherent peaks occur as a consequence of the rare extreme amplification of a single spontaneous emission event in the random laser. Clearly, this simple but interesting system is a testbed for the behavior of extreme events and their statistics.

4.6 Conclusion

Random lasers have always ignited academic curiosity among researchers, owing to the myriad statistical features they possess. They are capable of offering coherent emission, are compact in size, easy to fabricate, and can emit over a wide spectral range. The strong interactions between the multitude of modes realize the strong intensity and frequency fluctuations, which make them ideal candidates for investigations of statistical processes. Any understanding of the nature of fluctuations begins with a robust quantification scheme based on the systemic parameters. That is the part played by the tail exponent α of an α-stable law for the non-Gaussian distribution of the intensity statistics. The provision of this quantification was responsible for the identification of the three distinct statistical regimes: pre-lasing Gaussian, lasing Lévy, and self-averaged lasing Gaussian. The random laser system offers several finely tunable parameters, such as the pump energy, the disorder strength, and the sample size, each one of which can drive the system in and out of a Lévy regime. Based on the underlying physics of the random laser, the Gaussian–Lévy crossover serves as a universal identifier for the criticality in the system, and hence the random lasing threshold. This crossover essentially depends upon the gain length, and the mean path length of the diffusing photons, and so could be accurately traced using the diffusion equation, in an amplifying environment. All the three regimes of statistics could be constructed through a model of exponentially tempered Lévy sums. This model reproduces the strong interplay between various mechanisms in this complex system that determine the appearance of coherent peaks in the emission. The model reveals a perfect correlation between the rare extreme events of amplification and the occurrence of coherent peaks in the emission. Getting a fulsome ensemble on extremes is usually a challenging task, but the random laser provides a table-top system to generate multitudes of extremes, thus facilitating further studies of extremal statistics.

In closing, we note that the interesting statistical phenomena discussed in this chapter, such as non-Gaussian statistics, truncated Lévy sums, effects of extreme values, etc., add a new facet to random lasers as a promising testbed for various theoretically proposed phenomena in statistical physics.

References

Ambartsumyan, R. V., Basov, N. G., Kryukov, P. G., and Letokhov, V. S. (1966). Laser with non-resonant feedback, *JETP Letters* **3**, p. 167.

Ambartsumyan, R. V., Basov, N. G., Kryukov, P. G., and Letokhov, V. S. (1970). Non-resonant feedback in lasers, in J. H. Sanders and K. W. H. Stevens (eds.), *Progress in Quantum Electronics* **1**, pp. 107–185.

Anderson, P. W. (1958). Absence of diffusion in certain random lattices, *Physical Review Letters* **109**, p. 1492.

Balachandran, R. M., and Lawandy, N. M. (1996). Laser action in strongly scattering media, *Optics Letters* **21**, p. 1603.

Baudouin, Q., Mercadier, N., Guarrera, V., Guerin, W., and Kaiser, R. (2013). A cold-atom random laser, *Nature Physics* **9**, p. 357.

Berger, G. A., Kempe, M., and Genack, A. Z. (1997). Dynamics of stimulated emission from random media, *Physical Review E* **56**, p. 6118.

Bruno, R., Sorriso-Valvo, L., Carbone, V., and Bavassano, B. (2004). A possible truncated-Lévy-flight statistics recovered from interplanetary solar-wind velocity and magnetic-field fluctuations, *Europhysics Letters* **66**, p. 146.

Burroughs, S. M., and Tebbens, S. F. (2001). Upper-truncated power laws in natural systems, *Pure and Applied Geophysics* **158**, p. 741.

Cao, H., Xu, J. Y., Seelig, E. W., and Chang, R. P. H. (2000). Microlaser made of disordered media, *Applied Physics Letters* **76**, p. 2997.

Chakrabarty, A., and Meerschaert, M. M. (2011). Tempered stable laws as random walk limits, *Statistics and Probability Letters* **81**, p. 989.

del Castillo-Negrete, D. (2009). Truncation effects in superdiffusive front propagation with Lévy flights, *Physical Review E* **79**, p. 031120.

Fallert, J., Deitz, R. J. B., Sartor, J., Schneider, D., Klingshirn, C., and Kalt, H. (2009). Co-existence of strongly and weakly localized random laser mode, *Nature Photonics* **3**, p. 279.

Florescu, L., and John, S. (2004). Lasing in random amplifying medium: Spatiotemporal characteristics and nonadiabatic atomic dynamics, *Physical Review E* **70**, p. 036607.

Gajda, J., and Magdziarz, M. (2010). Fractional Fokker–Planck equation with tempered α-stable waiting times: Langevin picture and computer simulation, *Physical Review E* **82**, p. 011117.

Gnedenko, B. V., and Kolmogorov, A. N. (1954). *Limit Distribution for Sums of Independent Random Variables* (Addison-Wesley, Readin, MA).

Henyey, L. C., and Greenstein, J. L. (1941). Diffuse radiation in the galaxy, *Astrophysical Journal* **93**, p. 70.

Irvine, W. M. (1965). Multiple scattering by large particles, *Astrophysical Journal* **142**, p. 1563.

John, S., and Pang, G. (1996). Theory of lasing in a multiple-scattering medium, *Physical Review A* **54**, p. 3642.

Koutrouvelis, I. A. (1980). Regression-type estimation of the parameters of the stable laws, *Journal of American Statistical Association* **75**, p. 918.

Lawandy, N. M., Balachandran, R. M., Gomes, A. S. L., and Sauvain, E. (1994). Laser action in strongly scattering medi, *Nature* **368**, p. 436.

Lepri, S., Cavalieri, S., Oppo, G.-L., and Wiersma, D. S. (2007). Statistical regimes of random laser fluctuations, *Physical Review A* **75**, p. 063820.

Letokhov, V. S. (1968). Light generation by a scattering medium with a negative resonant absorption, *Soviet Physics JETP* **26**, p. 835.

Liu, J., Garcéa, P. D., Ek, S., Gregersen, N., Suhr, T., Schubert, M., Mørk, J., Stobbe, S., and Lodahl, P. (2014). Random nanolasing in the Anderson localized regime, *Nature Nanotechnology* **9**, p. 285.

Lévy, P. (1937). *Théorie de l'addition des variables aléatoires* (Jacques Gabay, Paris).

Mantegna, R. N., and Stanley, H. E. (1994). Stochastic processes with ultraslow convergence to a Gaussian: The truncated Lévy flight, *Physical Review Letters* **73**, p. 2946.

Mantegna, R. N., and Stanley, H. E. (2000). *An Introduction to Econophysics: Correlations and Complexity in Finance* (Cambridge University Press, UK).

Markushev, V. M., Zolin, V. F., and Briskina, C. M. (1986). Luminescence and stimulated emission of neodymium in sodium lanthanum molybdate powders, *Soviet Journal of Quantum Electronics* **16**, p. 281.

Matacz, A. (2000). Financial modeling and option theory with the truncated Lévy process, *International Journal of Theoretical and Applied Finance* **3**, p. 143.

Meerschaert, M. M., Roy, P., and Shao, Q. (2012). Parameter estimation for exponentially tempered power law distributions, *Communications in Statistics: Theory and Methods* **41**, p. 1839.

Meerschaert, M. M., Zhang, Y., and Baeumer, B. (2008). Tempered anomalous diffusion in heterogeneous systems, *Geophysical Research Letters* **35**, p. L17403.

Mujumdar, S., and Ramachandran, H. (2000). Spectral features in emission from random amplifying media, *Optics Communications* **176**, p. 31.

Mujumdar, S., Ricci, M., Torre, R., and Wiersma, D. S. (2004). Amplified extended modes in random lasers, *Physical Review Letters* **93**, p. 053903.

Nolan, J. P. (2013). Numerical calculations of stable densities and distribution functions, *Communications in Statistics: Stochastic Models* **13**, p. 759.

Oshanin, G., Holovatch, Y., and Schehr, G. (2011). Proportionate vs disproportionate distribution of wealth of two individuals in a tempered paretian ensemble, *Physica A* **390**, p. 4340.

Polson, R. C., and Vardeny, Z. V. (2004). Random lasing in human tissues, *Applied Physics Letters* **85**, p. 1289.

Redner, S. (2001). *A Guide to First-Passage Processes* (Cambridge: University Press, UK).

Rosinski, J. (2007). Tempering stable processes, *Stochastic Processes and their Applications* **117**, p. 667.

Samarodnitsky, G., and Taqqu, M. (1994). *Stable Non-Gaussian Random Processes: Stochastic Models with Infinite Variance* (Chapman and Hall, New York).

Sharma, D., Ramachandran, H., and Kumar, N. (2006a). Lévy statistical fluctuations from a random amplifying medium, *Fluctuations and Noise Letters* **6**, p. L95.

Sharma, D., Ramachandran, H., and Kumar, N. (2006b). Lévy statistics of emission from a novel random amplifying medium: An optical realization of the Arrhenius cascade, *Optics Letters* **31**, pp. 1806–1808.

Shlesinger, M. F., Zaslavsky, G. M., and Klafter, J. (1993). Strange kinetics, *Nature* **363**, p. 31.

Sibatov, R. T., and Uchaikin, V. V. (2011). Truncated Lévy statistics for dispersive transport in disordered semiconductors, *Communications in Nonlinear Science and Numerical Simulation* **16**, p. 4564.

Stanislavsky, A., Weron, K., and Weron, A. (2008). Diffusion and relaxation controlled by tempered α-stable processes, *Physical Review E* **78**, p. 051106.

Tiwari, A. K., and Mujumdar, S. (2013). Random lasing over gap states from a quasi-one-dimensional amplifying periodic-on- average random supperlattice, *Physical Review Letters* **111**, p. 233903.

Tiwari, A. K., Uppu, R., and Mujumdar, S. (2012). Aerosol-based coherent random laser, *Optics Letters* **37**, p. 1053.

Toublanc, D. (1996). Henyey–Greenstein and Mie phase functions in Monte Carlo radiative transfer computations, *Applied Optics* **35**, p. 3270.

Tulek, A., Polson, R. C., and Vardeny, Z. V. (2010). Naturally occuring resonators in random lasing of π-conjugated polymer films, *Nature Photonics* **6**, p. 303.

Turitsyn, S. K., Babin, S. A., El-taher, A. E., Harper, P., Churkin, D. V., Kablukov, S. I., Ania-casta nón, J. D., Karalekas, V., and Podivilov, E. V. (2010). Random distributed feedback fibre laser, *Nature Photonics* **4**, p. 231.

Uppu, R., and Mujumdar, S. (2010). Statistical uctuations of coherent and incoherent intensity in random lasers with nonresonant feedback, *Optics Letters* **35**, p. 2831.

Uppu, R., and Mujumdar, S. (2011). On the coherent modes of ultranarrow-band random lasers with nonresonant feedback, *Applied Optics* **50**, p. E13.

Uppu, R., and Mujumdar, S. (2013). Dependence of the Gaussian–Lévy transition on the disorder strength in random lasers, *Physical Review A* **87**, p. 013822.

Uppu, R., Tiwari, A. K., and Mujumdar, S. (2012). Identification of statistical regimes and crossovers in coherent random laser emission, *Optics Letters* **37**, p. 662.

van Albada, M. P., and Lagendijk, A. (1985). Observation of weak localization of light in a random medium, *Physical Review Letters* **55**, p. 2692.

van der Molen, K. L., Mosk, A. P., and Lagendijk, A. (2006). Intrinsic intensity fluctuations in random lasers, *Physical Review A* **74**, p. 053808.

Wiersma, D. S. (2008). The physics and applications of random lasers, *Nature Physic* **4**, p. 359.

Wiersma, D. S., and Lagendijk, A. (1996). Light diffusion with gain and random lasers, *Physical Review E* **54**, p. 4256.

Wolf, P.-E., and Maret, G. (1985). Weak localization and coherent backscattering of photons in disordered media, *Physical Review Letters* **55**, p. 2696.

Zhu, G., Gu, L., and Noginov, M. A. (2012). Experimental study of instability in a random laser with immobile scatterers, *Physical Review A* **85**, p. 043801.

Chapter 5

Spin-Glass Behavior of Random Lasers: Theory

Ernesto P. Raposo
Laboratório de Física Teórica e Computacional, Departamento de Física, Universidade Federal de Pernambuco, 50670-901 Recife, Pernambuco, Brazil
ernesto.raposo@ufpe.br

5.1 Introduction

In a series of pioneering works [1–10] by Angelani, Antennucci, Conti, Leuzzi, and collaborators, published in the last 15 years, seminal concepts and ideas, such as glassy phase transitions and replica symmetry breaking (RSB), borrowed from the original context of the statistical physics theory of disordered magnetic systems [11], were introduced in the study of photonic systems with disordered nonlinear medium, such as random lasers (RLs). Our aim in this chapter is to provide a brief review of these important theoretical achievements. Here we shall mainly be based on the description reported in the articles [5–7]. In particular, Ref. [7] is a highly recommended review on the statistical physics approach to

Lévy Statistics and Spin Glass Behavior in Random Lasers
Edited by Anderson S. L. Gomes, André L. Moura, Cid B. de Araújo, and Ernesto P. Raposo

ISBN 978-981-4968-55-3 (Hardcover), 978-1-003-33618-1 (eBook)
www.jennystanford.com

RLs, where the interested reader will find further technical details. Our notation will follow as much as close to that of Refs. [5–7].

The concept of RSB was first presented [11–13] by Parisi in the late 1970s in the study of the magnetic properties of strongly disordered spins systems. In particular, it was originally applied in the context of the spin-glass phase, in which the spins "freeze" in random directions at low temperatures. In this glassy phase, spins present strong correlations in time and slow dynamics fluctuations, if compared with the essentially uncorrelated replica-symmetric paramagnetic phase, which displays rapidly fluctuating spins at high temperatures.

In that context, the concept of RSB was applied to justify that an identically prepared spin-glass system with the same distribution of randomly disordered long-range spin-spin interactions under identical conditions (i.e., copies or replicas of the spin system) can reach different states and lead to distinct measurements of observables. The RSB phenomenon emerges as the system's free energy landscape breaks into a large number of local minima in the configuration space, which are separated by relevant energy barriers. Since the system can become eventually trapped in one of these local minima for a considerably long time, thus hampering its ergodic behavior, properties dependent on the history of the measurement arise, such as irreversibility effects and magnetic hysteresis.

The characterization of the RSB phenomenon in magnetic spin systems can be performed through the analysis of the distribution of values of a spin-spin correlation function for a large number of system replicas [11–13]. This quantity measures the overlap between every two replicas of the system, being thus termed the replica overlap parameter (see below).

Essentially, if the distribution of the values of this replica overlap parameter is centered around zero, then replicas are considered symmetric (independent, uncorrelated), a scenario observed in the paramagnetic phase at high temperatures [11]. Replica symmetry can be also identified at the ferromagnetic state without the disorder below the critical temperature.

On the other hand, if the distribution of replica overlap values displays a maximum (or maxima) at non-zero values of the replica overlap parameter, then the symmetry of replicas is broken and a spin-glass phase with RSB and nontrivial correlations among system replicas emerges in these strongly disordered magnetic systems at temperatures below some critical freezing temperature [11].

The spin-glass state is magnetically frustrated in the sense no configuration of spins can simultaneously satisfy all random interactions among them. Disorder and frustration are then considered to be essential ingredients for a spin-glass phase with RSB [11]. Moreover, due to the random orientation of spins the system's magnetization is null in the low-temperature spin-glass phase, just as in the high-temperature paramagnetic state. Therefore, magnetization cannot be considered as the order parameter of the spin-glass phase. Instead, the order parameter of this phase is associated with the distribution of values of the replica overlap parameter [11–13], since the critical freezing temperature marks the frontier between phases with quite distinct behaviors of this quantity.

These ideas inspired the authors of Refs. [1–10] in the analysis of the Hamiltonian-like formulation of open and close multimode photonic systems with disorder and nonlinearity. As a consequence, a rich phase diagram of such systems was built in terms of the disorder strength, nonlinearity degree, and input excitation power, which includes both replica-symmetric and RSB phases, as reviewed in the following.

5.2 Hamiltonian-Like Formulation of RL Systems

The Hamiltonian-like formulation of open or closed multimode photonic systems with disorder or not can be traced back to the works of Hackenbroich, Viviescas, and Haake [14–16], which, in turn, have found inspiration from the system-and-bath Hamiltonian models of open quantum systems, including nonlinear optical ones [17].

We begin by briefly reviewing the derivation [14–16] of the effective Hamiltonian of open multimode laser systems with disorder, which also comprises RLs and random fiber lasers. The quantum system-and-bath Hamiltonian of such systems can be written in the general form:

$$\mathcal{H} = \mathcal{H}_{\mathrm{S}} + \mathcal{H}_{\mathrm{B}} + \mathcal{H}_{\mathrm{SB}}, \tag{5.1}$$

where the first two terms above describe, respectively, the quantized electromagnetic field inside and outside (i.e., the bath region) the system, while $\mathcal{H}_{\mathrm{SB}}$ accounts for the system-bath interaction.

Starting from the classical electromagnetic energy, the Feshbach projection technique can be applied [14–16] to separate the contributions to the electromagnetic field from the regions inside and outside the system, which after proper quantization of the fields allow to identify (in $\hbar$ units)

$$\begin{aligned}
\mathcal{H}_{\mathrm{S}} &= \sum_k \omega_k \alpha_k^\dagger(t)\alpha_k(t), \\
\mathcal{H}_{\mathrm{B}} &= \int d\omega \sum_m \omega \beta_m^\dagger(\omega, t)\beta_m(\omega, t), \\
\mathcal{H}_{\mathrm{SB}} &= \int d\omega \sum_{k,m} \mathcal{W}_{km}(\omega)\alpha_k^\dagger(t)\beta_m(\omega, t) + \text{h.c.},
\end{aligned} \tag{5.2}$$

where $\alpha_k^\dagger(t)$ and $\alpha_k(t)$ are creation and annihilation operators obeying canonical commutation relations in the Heisenberg representation, which are assigned to the field eigenmodes of the system with discrete eigenfrequencies ω_k indexed by integers k. Analogously, $\beta_m^\dagger(\omega, t)$ and $\beta_m(\omega, t)$ form a continuous set of operators for the modes outside the system with continuous frequency ω and channel discrete index m related, e.g., to the expansion of the outer solutions in angular momentum states [14–16]. Thus, $\mathcal{H}_{\mathrm{B}}$ describes a large reservoir (or external thermal bath) of quantum oscillators interacting bilinearly with the $\alpha_k^\dagger(t)$ and $\alpha_k(t)$ operators, as indicated in $\mathcal{H}_{\mathrm{SB}}$ by the coupling amplitudes $\mathcal{W}_{km}(\omega)$ in the rotating-wave approximation [14–16]. Clearly, the description of closed systems, with no energy leakage to the outer region and no interaction with the external bath, involves only $\mathcal{H}_{\mathrm{S}}$.

The Heisenberg equations of motion that follow from Eqs. (5.1) and (5.2) read

$$\frac{d\alpha_k}{dt} = -i\omega_k\alpha_k - i\int d\omega \sum_m \mathcal{W}_{km}\beta_m \tag{5.3}$$

and

$$\frac{d\beta_m}{dt} = -i\omega\beta_m - i\sum_k \hbar\mathcal{W}^*_{km}\alpha_k. \tag{5.4}$$

We observe that the dynamics equation for the fields $\beta_m(\omega, t)$ can be formally solved, with the result substituted into Eq. (5.3) in the treatment below.

Moreover, in order to account for the nonlinear effects arising from the gain in the system, the Jaynes–Cummings atom-field interaction has to be added to $\mathcal{H}_S$ in Eq. (5.2) [18]. This implies the introduction of the term $\int d\mathbf{r} g^*_k(\mathbf{r})\sigma_-(\mathbf{r}, t)$ in the right-hand side of the $d\alpha_k/dt$ dynamics in Eq. (5.3), with the integral performed over the inner region and σ_- denoting the atomic density lowering operator of the atoms (or ions) responsible for the stimulated emissions of photons. The atom-field coupling $g_k(\mathbf{r})$ is proportional to $\mathbf{p} \cdot \mathbf{E}_k(\mathbf{r})$, where $\mathbf{p}$ represents the atomic dipole matrix elements and $\mathbf{E}_k(\mathbf{r})$ is the electric field amplitude inside the system. In the case of a system's nonlinear medium with some kind of structural disorder, its refractive index $n(\mathbf{r})$ displays a random spatial profile, which particularly influences the behavior of the electromagnetic field.

With the introduction of the atom-field interaction, the time evolution of the operators α_k and β_m have to be considered along with the dynamics equations for σ_- and the density inversion operator $\sigma_z(\mathbf{r}, t)$, namely [15, 16],

$$\frac{d\sigma_-}{dt} = -(\gamma_\perp + i\omega_a)\sigma_- + 2\sum_k g_k\sigma_z\alpha_k + F_-, \tag{5.5}$$

and

$$\frac{d\sigma_z}{dt} = \gamma_\parallel(S\rho - \sigma_z) - \sum_k (g^*_k\alpha^\dagger_k\sigma_- + \text{h.c.}) + F_z, \tag{5.6}$$

where $\gamma_\perp$ and $\gamma_\parallel$ denote the decay rates, ω_a indicates the frequency of the active atomic transition, $\rho(\mathbf{r})$ is the atomic density, the noise terms represented by $F_-(\mathbf{r}, t)$ and $F_z(\mathbf{r}, t)$ arise from the coupling of the system's atoms with the external bath, and S is the pump intensity.

Since the dynamics of σ_- and σ_z evolve much faster than that of α_k and β_m, then Eqs. (5.5) and (5.6) can be solved recursively in the stationary regime and, in the present approach, the solution for σ_- up to the third order in the fields α_k, neglecting the noise terms in Eqs. (5.5) and (5.6), is substituted back into the equation of motion for $d\alpha_k/dt$. Further, by considering [14–16] coupling amplitudes $\mathcal{W}_{km}$ independent of ω for a sufficiently large band around ω_a (Markov approximation), the final form of $d\alpha_k/dt$ is compared with the Heisenberg equation of motion for $\alpha_k(t)$ thus allowing to write [14–16] the effective quantum non-Hermitian (due to the openness of the system) Hamiltonian-like expression for the multimode photonic system up to the fourth-order in the operators α_k and $\alpha_k^\dagger$,

$$\mathcal{H}_S = -\frac{1}{2}\sum_{k_1 k_2} g^{(2)}_{k_1 k_2}\alpha^\dagger_{k_1}\alpha_{k_2} - \frac{1}{4!}\sum_{k_1 k_2 k_3 k_4} g^{(4)}_{k_1 k_2 k_3 k_4}\alpha^\dagger_{k_1}\alpha_{k_2}\alpha^\dagger_{k_3}\alpha_{k_4}, \tag{5.7}$$

with

$$\begin{aligned} g^{(2)}_{k_1 k_2} &= -2\omega_{k_1}\delta_{k_1 k_2} + 2\pi i(\mathcal{W}\mathcal{W}^\dagger)_{k_1 k_2} + G^{(2)}_{k_1 k_2}, \\ G^{(2)}_{k_1 k_2} &\propto S\int d\mathbf{r}\rho(\mathbf{r})g^*_{k_1}(\mathbf{r})g_{k_2}(\mathbf{r}), \\ g^{(4)}_{k_1 k_2 k_3 k_4} &\propto S\int d\mathbf{r}\rho(\mathbf{r})g^*_{k_1}(\mathbf{r})g_{k_2}(\mathbf{r})g^*_{k_3}(\mathbf{r})g_{k_4}(\mathbf{r}). \end{aligned} \tag{5.8}$$

Likewise, the system-bath interaction Hamiltonian in Eq. (5.2) can be written in the form:

$$\mathcal{H}_{SB} = \sum_k iF_k\alpha^\dagger_k + \text{h.c.}, \tag{5.9}$$

with

$$F_k(t) \equiv -i\int d\omega\sum_m e^{-i\omega(t-t_0)}\mathcal{W}_{km}\beta_m(\omega, t_0) \tag{5.10}$$

interpreted as a quantum noise operator [19] that depends on the bath operators only at the given time $t = t_0$ [14–16]. From Eq. (5.10), we observe that the noise correlation $\langle F^*_k(t)F_{k'}(t')\rangle$ is in general not proportional to $\delta_{k,k'}\delta(t-t')$, thus implying a non-white noise. However, the noise can be made [5–7] white and uncorrelated in a suitable basis of slow-amplitude modes in the semiclassical context, as discussed below.

In the present approach, the quartic coupling $g^{(4)}$ in the system's effective Hamiltonian, Eq. (5.7), is the main source of nonlinearity, as it relates to the nonlinear susceptibility $\chi^{(3)}$. We further remark that the disordered character of the couplings $g^{(p)}$ in $\mathcal{H}_S$, both quadratic (order $p = 2$) and quartic ($p = 4$), is due in part to the random positions of the stimulated atoms or ions and the random spatial profile of the refractive index.

Noticeably, a similar quantum system-and-bath Hamiltonian of interacting spins, also expressed in terms of creation and annihilation operators, was considered in Ref. [20], in which the external bath of quantum oscillators couples as well in a bilinear form with the system's spin operators. In this case, the general effective quantum Hamiltonian displays spins interacting up to the general p-th order. Also, a spherical constraint on the sum of squared spin operators is introduced [20], thus defining a quantum p-spin spherical model (in the photonic context, the analogous constraint corresponds to fixing the total optical power [5–7], see below). Remarkably, when random p-th order couplings with Gaussian distributions are considered in the case $p > 2$, a spin-glass phase emerges [20] under Parisi's RSB formalism [11]. In the next section, we shall see that these features will find their photonic counterparts in the statistical physics analysis of RL systems governed by an effective Hamiltonian analogous to Eq. (5.7).

5.3 Statistical Physics Approach to RL Systems

In the semiclassical context, a Hamiltonian-like formulation similar to that of Eq. (5.7), but with the quantum operators $\alpha_k^\dagger$ and α_k replaced by their complex expectation values (α_k^* and α_k, respectively), also yields a photonic RSB spin-glass phase in multimode RL systems in closed or open cavities for sufficiently strong disorder in the $p = 2$ and $p = 4$ random couplings [5–7].

One further observes that, in addition to considering the expectation values of the field operators in the semiclassical version of Eq. (5.7), it is also convenient to work [5–7] on a basis of so-called slow-amplitude modes $\bar{a}_k(t)$, defined as:

$$\alpha_\lambda(t) = \sum_k A_{\lambda k} \bar{a}_k(t), \tag{5.11}$$

with

$$\bar{a}_k(t) = a_k(t)\exp(-i\omega_k), \tag{5.12}$$

where the dynamics of the modes amplitudes $a_k(t)$ evolve much slower than the typical time scales ω_k^{-1}. At this point, we note that, since the choice of basis in Eq. (5.11) is not unique, one can pick [5–7] a basis of slow-amplitude modes in which the noise F_λ, Eq. (5.10), due to coupling to the external bath is white and uncorrelated.

From Eq. (5.7), and neglecting the semiclassical form of Eq. (5.9) for the noise term related to the system-bath interaction, the semiclassical effective Hamiltonian-like of the multimode laser system is given by

$$\mathcal{H} = -\frac{1}{2}\sum_{k_1k_2|\mathrm{FMC}} \bar{g}^{(2)}_{k_1k_2} a^*_{k_1} a_{k_2} - \frac{1}{4!}\sum_{k_1k_2k_3k_4|\mathrm{FMC}} \bar{g}^{(4)}_{k_1k_2k_3k_4} a^*_{k_1} a_{k_2} a^*_{k_3} a_{k_4}, \tag{5.13}$$

in which the bar over the quadratic and quartic couplings denotes that the matrix elements $A_{\lambda k}$ have been incorporated with the change of basis (Eq. (5.11)), and FMC in the sums indicates that the frequency matching conditions apply, that is, $|\omega_{k_1} - \omega_{k_2}| \lesssim \gamma$ in the first sum and $|\omega_{k_1} - \omega_{k_2} + \omega_{k_3} - \omega_{k_4}| \lesssim \gamma$ in the second one, with γ representing the typical mode linewidth. These constraints arise from the Fourier transform of the slow amplitude modes, in the form $\bar{a}_k(\omega) \approx \delta(\omega - \omega_k)$. Hamiltonian Eq. (5.13) thus represents the semiclassical picture in the approximation of slow amplitude modes for the description of the photonic properties of closed or open systems with disordered nonlinear media. In particular, we recall that RLs are generally open systems with nonlinear random media.

As discussed, the quadratic and quartic couplings in Eq. (5.13) carry the signature of the structural disorder in the case of RL systems. Their explicit calculation from the first principles is virtually unfeasible. Therefore, for practical purposes, one possibility is to assume that their values are taken from some probability density function (PDF) [5–7].

By considering in the mean-field approach that the couplings between modes are statistically independent, with the first and second moments of their distributions actually being the only essential ones, and further assuming that all couplings in a given (quadratic or quartic) order are taken from the same PDF, along with the relaxation of the FMC constraints, so that any given combination of modes is allowed in each sum, one can thus recast Eq. (5.13) in the Hamiltonian form [5–7],

$$\mathcal{H} = -\frac{1}{2}\sum_{jk} J_{jk} a_j^* a_k - \frac{1}{4!}\sum_{jklm} J_{jklm} a_j^* a_k a_l^* a_m, \tag{5.14}$$

in which the real-valued second- and fourth-order couplings, respectively denoted by J_{jk} ($p = 2$) and J_{jklm} ($p = 4$), are given [5–7] in the mean-field approach by Gaussian distributions generally written as:

$$P(J_{i_1 \ldots i_p}) = \frac{1}{\sqrt{2\pi\sigma_p^2}} \exp\left[-\frac{\left(J_{i_1 \ldots i_p} - \bar{J}_0^{(p)}\right)^2}{2\sigma_p^2}\right]. \tag{5.15}$$

Above we notice that the first and second moments of the $p = 2$ and $p = 4$ Gaussian PDFs have been conveniently rescaled by the number N of the system's modes,

$$\sigma_p^2 = \frac{p! J_p^2}{2N^{p-1}} \quad \text{and} \quad \bar{J}_0^{(p)} = \frac{J_0^{(p)}}{N^{p-1}}, \tag{5.16}$$

where $J_0^{(p)}$ and J_p are energy scales respectively associated with the disordered (variance) and ordered (mean) properties of the system. In Refs. [5–7], the authors introduce the following parametrization,

$$J_0^{(4)} = \alpha_0 J_0, \quad \alpha_0 = \left(\frac{J_0^{(2)}}{J_0^{(4)}} + 1\right)^{-1}, \quad J_0 = J_0^{(2)} + J_0^{(4)},$$

$$J_4 = \alpha J, \quad \alpha = \left(\frac{J_2}{J_4} + 1\right)^{-1}, \quad J = J_2 + J_4, \tag{5.17}$$

in which the degree of nonlinearity is controlled by the parameters α_0 and α, with higher values in the range [0, 1] meaning stronger nonlinear character. In addition, for a system in equilibrium with the external pump source, the total optical intensity is a constant,

$$\mathcal{E} = \epsilon N = \sum_k |a_k|^2, \tag{5.18}$$

with the parameter ϵ essentially representing the mean intensity per mode. As mentioned above, this constraint is analogous to that of spherical spin models [20, 21], and so Eq. (5.14) can be in fact seen as a complex (due to the complex nature of the variables a_k) $(p = 2) + (p = 4)$ spherical semiclassical Hamiltonian for generally open and disordered multimode laser systems. Such resemblance also opens up the possibility for a photonic spin-glass phase with RSB properties in RL systems, as detailed below, in analogy to the one found in the case of $(p = 2)+(p = 4)$ disordered spherical spin models [20, 21].

The strength of the disorder and pumping rate can be expressed, respectively, by

$$R_J = \frac{J}{J_0} \quad \text{and} \quad \mathcal{P} = \epsilon\sqrt{\beta J_0}, \tag{5.19}$$

in which $\beta = (k_B T)^{-1}$, k_B denotes the Boltzmann constant, and T represents the heat-bath temperature (proportional to the white noise self-correlation). It is important to notice from Eq. (5.19) that the photonic temperature essentially plays the role of the inverse input excitation power. In other words, it is thus possible to anticipate that the low-temperature phases in the analogous description of disordered magnetic systems (such as the spin-glass one with RSB) will eventually find their counterparts in the high excitation-power photonic regimes above the laser threshold in RLs. In the above parametrization, the system's phase diagram arising from the statistical physics analysis of the Hamiltonian Eq. (5.14) will be plotted below as a function of the relevant parameters that determine its photonic behavior, namely, $\alpha = \alpha_0$, R_J, and $\mathcal{P}$.

To proceed further, one can separate the real and imaginary parts of the complex modes' amplitude variables a_k in the rescaled form [5–7],

$$a_k = \sqrt{\epsilon}(\sigma_k + i\tau_k), \tag{5.20}$$

so that the spherical constant on the total optical intensity, Eq. (5.18), reads

$$\sum_k (\sigma_k^2 + \tau_k^2) = N. \tag{5.21}$$

Now, by substituting Eq. (5.20) into Eq. (5.14), the Hamiltonian $\mathcal{H}$ can be conveniently recast in terms of the new set of real variables σ_k and τ_k.

The statistical physics approach to Eq. (5.14) starts by writing the partition function of the photonic system for a given realization J of the Gaussian second- and fourth-order couplings between modes as

$$\mathcal{Z}_N[J] = \int \mathcal{D}a \exp\big(-\beta\mathcal{H}[a, J]\big), \tag{5.22}$$

with

$$\mathcal{D}a \equiv \prod_{k=1}^{N} da_k da_k^*. \tag{5.23}$$

The free energy per number N of modes in the thermodynamic limit is expressed by properly weighting each realization J in the form

$$f = -\lim_{N\to\infty} \frac{1}{\beta N} \int \mathcal{D}J\, P[J] \ln \mathcal{Z}_N[J] \equiv -\lim_{N\to\infty} \frac{1}{\beta N} \overline{\ln \mathcal{Z}_N[J]}. \tag{5.24}$$

At this point, we comment that the calculation of the free energy density f can be performed with the help of the so-called replica trick [11], which takes advantage of a well-known limit form for the ln function in Eq. (5.24). In this approach, (integer) n copies of the system are considered with the same realization and distributions of couplings, and by continuing n to real the limit $n \to 0$ is taken so that

$$f = -\lim_{N\to\infty} \lim_{n\to 0} \frac{1}{\beta N} \left[\frac{\overline{(\mathcal{Z}_N[J])^n} - 1}{n} \right], \tag{5.25}$$

where

$$\overline{(\mathcal{Z}_N[J])^n} = \int \mathcal{D}J\, P[J] \int \mathcal{D}a_1 \ldots \mathcal{D}a_n \exp\Big[-\beta\big(\mathcal{H}[a_1, J] + \ldots + \mathcal{H}[a_n, J]\big)\Big]. \tag{5.26}$$

Notice that the exponential in Eq. (5.26) presents a sum of Hamiltonians (Eq. (5.14)), each one of them with the same realization J of second- and fourth-order couplings between modes, but with distinct sets of modes amplitude variables, $a_a \equiv \{a_k^a\}$, with mode index $k = 1, 2, \ldots, N$, assigned to different system's replicas $a = 1, 2, \ldots, n$ (do not confound the replica index a with the modes amplitude variables a_k^a; replica indexes will be represented here by the letters a and b).

It is further convenient to define [5–7] the following matrices and variables,

$$Q_{ab} = \frac{1}{N}\sum_k (\sigma_k^a \sigma_k^b + \tau_k^a \tau_k^b) = \frac{1}{\epsilon}\sum_k \text{Re}\left[a_k^a (a_k^b)^*\right], \tag{5.27}$$

$$R_{ab} = \frac{1}{N}\sum_k (\sigma_k^a \sigma_k^b - \tau_k^a \tau_k^b) = \frac{1}{\epsilon}\sum_k \text{Re}\left[a_k^a a_k^b\right], \tag{5.28}$$

and

$$m_\sigma = \frac{\sqrt{2}}{N}\sum_k \sigma_k^a, \quad m_\tau = \frac{\sqrt{2}}{N}\sum_k \tau_k^a, \tag{5.29}$$

so that

$$m_\sigma + i m_\tau = \frac{\sqrt{2}}{\epsilon}\sum_k a_k^a. \tag{5.30}$$

We observe above that Q_{ab} and R_{ab} take into account the overlap between the system's replicas a and b, in a type of complex analog to the real replica overlap parameter in Parisi's RSB approach to the spin-glass theory of disordered magnetic systems [11]. Similarly, the variables m_σ and m_τ are the magnetization analogs. We also observe that the replica index a does not appear in m_σ and m_τ, as justified in the following. The quantities Q_{ab}, R_{ab}, m_σ, and m_τ actually constitute [5–7] the order parameters that will characterize the diverse photonic phases of the multimode laser system.

In terms of the quantities defined in Eqs. (5.27)–(5.29), the n-th order average of the partition function in Eq. (5.26) is expressed as [5–7]:

$$\overline{\mathcal{Z}_N^n} = \int \mathcal{D}Q\,\mathcal{D}R\,\mathcal{D}m_\sigma\,\mathcal{D}m_\tau \,\exp\big(-NG[Q, R, m_\sigma, m_\tau]\big), \tag{5.31}$$

in which the functional G reads

$$\begin{aligned} G[Q, R, m_\sigma, m_\tau] = &-\frac{1}{2}\sum_{ab} g(Q_{ab}, R_{ab}) - nk(m_\sigma, m_\tau) \\ &-\frac{1}{2}\ln\det(Q+R) + \frac{m_\sigma^2}{2}\sum_{ab}(Q+R)_{ab}^{-1} \\ &-\frac{1}{2}\ln\det(Q-R) + \frac{m_\tau^2}{2}\sum_{ab}(Q-R)_{ab}^{-1}, \end{aligned} \tag{5.32}$$

with the following definition of auxiliary functions,

$$g(x, y) = \xi_2(x^2 + y^2) + \frac{\xi_4}{2}(x^4 + y^4 + 4x^2y^2),$$
$$k(x, y) = b_2(x^2 + y^2) + b_4(x^2 + y^2)^4. \qquad (5.33)$$

Above parameters b_2, b_4, ξ_2, and ξ_4 can be expressed in terms of the system's relevant variables α_0, α, R_J, and $\mathcal{P}$ [see Eqs. (5.17) and (5.19)] as

$$b_2 = \frac{(1-\alpha_0)}{4}\mathcal{P}\sqrt{\beta J_0}, \quad b_4 = \frac{\alpha_0}{96}\mathcal{P}^2,$$
$$\xi_2 = \frac{(1-\alpha)^2}{4}\beta J_0 \mathcal{P}^2 R_J^2, \quad \xi_4 = \frac{\alpha^2}{6}\mathcal{P}^4 R_J^2. \qquad (5.34)$$

In order to handle the integral in Eq. (5.31) in the replica space, the saddle-point extremization method can be applied in the thermodynamic limit, so that

$$f = f_0 + \lim_{n\to 0} \text{Extr}\, \frac{1}{\beta n} G[Q, R, m_\sigma, m_\tau], \qquad (5.35)$$

where the constant f_0 is thermodynamically irrelevant and G is evaluated at its stationary point in the limit $n \to 0$, i.e., a point of maximum (minimum) with respect to variations in Q and R (m_σ and m_τ) [5–7].

Indeed, the next step consists in the calculation of the functional $G[Q, R, m_\sigma, m_\tau]$, Eq. (5.32), at the stationary point. In this sense, by following the prescription described in [5–7], based on Parisi's RSB approach to spin-glass systems [11–13], one is first led to the conclusion that the simplest ansatz of symmetry of replicas, so that $Q_{ab} = Q$ and $R_{ab} = R$ for any pair of distinct replicas a and b, is not thermodynamically stable in the most relevant part of the parameters' space, notably the region with high-excitation powers (analog low temperatures). Notwithstanding, the variables m_σ and m_τ defined in Eq. (5.29) are replica symmetric (i.e., independent of the particular replica a) since they are single replica quantities [5–7].

A similar difficulty with the replica symmetric ansatz was faced by Parisi in the statistical approach to spin-glass systems in the late 1970s. To circumvent this problem, Parisi introduced the seminal idea of RSB [11–13], in which, depending on the region of the parameters' space, two systems replicas a and b can in principle not be fully equivalent or symmetric. This implies that the values of Q_{ab} may differ for $a \neq b$ (the same applies for R_{ab}). In this context, the probability distribution of the values of these quantities becomes an order parameter separating the phases with and without RSB. Moreover, in the case when the replica symmetry is broken, it may be possible that identically prepared systems evolving under equivalent thermodynamical conditions (i.e., copies of the system) reach different equilibrium configurations, thus giving rise to properties that are generally dependent on the history of the measurement, as discussed in the Introduction.

From a practical point of view, Parisi designed a calculation scheme [11–13] for the RSB approach in which, in the case of a system with n replicas, an $n \times n$ matrix M can be constructed in an $\mathcal{R}$-step RSB state described by $\mathcal{R} + 2$ parameters $(M_{\mathcal{R}+1}, M_{\mathcal{R}}, \ldots, M_0)$ [5–7]. Such matrix is then divided into diagonal matrix blocks of decreasing linear size p_r, with $1 = p_{\mathcal{R}+1} < p_{\mathcal{R}} < \ldots < p_0 = n$. Moreover, one further assigns $M_{ab} = M_r$ if the replicas a and b belong to the same block (see [11–13] and [5–7] for more details on Parisi's RSB calculation scheme).

Before continuing, it is interesting to notice in Eqs. (5.27) and (5.28) that both matrices Q_{ab} and R_{ab} comprise mixtures of products $\sigma_k^a \sigma_k^b$ and $\tau_k^a \tau_k^b$ of the real and imaginary parts of the modes' amplitude variables a_k^a and a_k^b, see Eq. (5.20). This can be readily disentangled by defining the following matrices to be employed in the subsequent calculations,

$$
\begin{aligned}
A_{ab} &= Q_{ab} + R_{ab} = \frac{2}{N} \sum_k \sigma_k^a \sigma_k^b, \\
B_{ab} &= Q_{ab} - R_{ab} = \frac{2}{N} \sum_k \tau_k^a \tau_k^b.
\end{aligned} \tag{5.36}
$$

The stationarity analysis of the functional $G[Q, R, m_\sigma, m_\tau]$ given in Eq. (5.32) with respect to variations in Q, R, m_σ, and m_τ is actually quite involving [5–7]. By expressing the results obtained in terms of the matrices defined in Eq. (5.36), one finds [5–7] the following saddle-point equations in the replica-trick limit $p_0 = n \to 0$,

$$\Lambda(Q_0, R_0) - \frac{(A_0 - m_\sigma^2)}{(A_{\widehat{1}}^2)} - \frac{(B_0 - m_\tau^2)}{(B_{\widehat{1}}^2)} = 0,$$
$$\Lambda(Q_r, R_r) - \Lambda(Q_{r-1}, R_{r-1}) - \frac{(A_r - A_{r-1})}{A_{\widehat{r}} A_{\widehat{r+1}})} - \frac{(B_r - B_{r-1})}{(B_{\widehat{r}} B_{\widehat{r+1}})} = 0, \tag{5.37}$$

for variations regarding Q_{ab}, with $r = 1, \ldots, \mathcal{R}$ and defining $\Lambda(x, t) = \partial g(x, y)/\partial x$,

$$\Lambda(R_0, Q_0) - \frac{(A_0 - m_\sigma^2)}{(A_{\widehat{1}}^2)} + \frac{(B_0 - m_\tau^2)}{(B_{\widehat{1}}^2)} = 0,$$
$$\Lambda(R_r, Q_r) - \Lambda(R_{r-1}, Q_{r-1}) - \frac{(A_r - A_{r-1})}{A_{\widehat{r}} A_{\widehat{r+1}})} + \frac{(B_r - B_{r-1})}{(B_{\widehat{r}} B_{\widehat{r+1}})} = 0, \tag{5.38}$$

for variations in R_{ab}, with $r = 1, \ldots, \mathcal{R}+1$ and the boundary values $Q_{\mathcal{R}+1} = 1$ and $R_{\mathcal{R}+1} = \bar{R}$, and finally

$$m_\sigma = b A_{\widehat{1}}, \tag{5.39}$$

for variations in m_σ or m_τ, with $b = (2b_2 + 4b_4 m_\sigma^2) m_\sigma$ and by considering the case $m_\tau = 0$ (from now on, $m_\tau = 0$ will be taken without loss of generality [5–7]). Above the hatted notation represents the replica Fourier transform of a $\mathcal{R}$-RSB matrix M_r [5–7], given by

$$M_{\widehat{k}} = \sum_{r=k}^{\mathcal{R}+1} p_r (M_r - M_{r-1}),$$
$$M_r = \sum_{k=0}^{r} \frac{1}{p_k} (M_{\widehat{k}} - M_{\widehat{k+1}}), \tag{5.40}$$

with null values for these quantities when the indices lie outside the respective intervals of definition. In this context, the saddle-point solution for G in the $\mathcal{R}$-RSB approach reads

$$G = -\frac{1}{2} \sum_{r=0}^{\mathcal{R}+1} (p + r - p_{r+1}) g(Q_r, R_r) - k(m_\sigma, m_\tau) + \frac{m_\sigma^2}{2A_{\widehat{0}}} + \frac{m_\tau^2}{2B_{\widehat{0}}}$$
$$-\frac{1}{2} \ln (A_{\mathcal{R}+1} - A_{\mathcal{R}}) - \frac{1}{2} \sum_{r=1}^{\mathcal{R}+1} \frac{1}{p_r} \ln \left(\frac{A_{\widehat{r}}}{A_{\widehat{r+1}}} \right) + \frac{A_0}{A_{\widehat{1}}}$$
$$-\frac{1}{2} \ln (B_{\mathcal{R}+1} - B_{\mathcal{R}}) - \frac{1}{2} \sum_{r=1}^{\mathcal{R}+1} \frac{1}{p_r} \ln \left(\frac{B_{\widehat{r}}}{B_{\widehat{r+1}}} \right) + \frac{B_0}{B_{\widehat{1}}}. \tag{5.41}$$

The stationarity equations, Eqs. (5.37)–(5.39), present [5–7] solutions with $m_\tau = 0$ in the general form $B_r = 0$ and $Q_r = R_r$, for $r = 0, \ldots, \mathcal{R}$.

5.4 Phase Diagram of Multimode Laser Systems

In Refs. [5–7], the authors employed the above calculation scheme in the specific case for $\mathcal{R} = 1$, so that $r = 0, 1$, in the so-called one-step RSB approach (1RSB) to the functional $G[Q, R, m_\sigma, m_\tau]$ in Eq. (5.32). In this context, they obtained [5–7] a remarkably rich photonic phase diagram as a function of the relevant variables $\alpha = \alpha_0$ (degree of nonlinearity), R_J (disorder strength), and $\mathcal{P}$ (pumping rate), with the phases characterized by the values of the order parameters Q_{ab}, R_{ab}, and m_σ (henceforth referred to simply as m).

Four phases have been identified in [5–7], as depicted in Figs. 5.1 and 5.2 and generally described in the following, including the indication of their equivalent regime in the analogous disordered spins systems.

- Incoherent wave (IW) (also termed continuous wave (CW) in [5]) for small pumping powers and any disorder and nonlinearity degrees. This is a phase with symmetry of replicas, in which all order parameters have null values. Modes oscillate independently without coherence, and the emitted light has the form of a continuous wave. It is analogous to the paramagnetic regime in spins systems with or without the disorder.
- Mode-locking laser (ML) (also termed standard mode-locking laser (SML) in [5, 6]) for pumping powers above some threshold value, disorder strength below some upper limit, and any nonlinearity degree. This phase can appear with or without the disorder, and present RSB or not, for non-null values of the magnetization order parameter, $m \neq 0$. The lasing modes oscillate coherently with the same phase, and the emitted light has the form of short pulses. It is analogous to the ferromagnetic (random bond

ferromagnetic) phase in spins systems without (with weak) disorder.

- RL for pumping powers above some threshold, disorder strength above some lower limit, and any nonlinearity degree. Glassy RSB regime with $m = 0$ and existence of non-null values of the order parameters Q_{ab} and R_{ab}, which present nontrivial structure. Modes are coherent in phase but not in intensity, with frustrated synchronous oscillation. It is analogous to the RSB spin-glass phase in disordered spins systems.
- Phase locking wave (PLW) for intermediate pumping powers in the regime between the IW and RL phases, with non-null disorder strength and any nonlinearity degree. Replicas are symmetric and the order parameters have null values, except for the diagonal elements R_{aa}, since $\langle\sigma^2\rangle \neq \langle\tau^2\rangle$, see Eq. (5.28). It has no analogous in spins systems.

Figures 5.1 and 5.2 show the phase diagrams in the parameters' space ($\mathcal{P}^2$, R_J, $\alpha = \alpha_0$), with focus on the cases of open ($\alpha = \alpha_0 < 1$) and closed ($\alpha = \alpha_0 = 1$, no cavity energy leakage and no interaction with the external bath) systems. All phases discussed above can be identified in these figures (we note that the phases' nomenclature of [5] has been used).

Interestingly, it was observed in [5–7] that low values of α_0 generally give rise to continuous second-order phase transitions in the phase diagrams of Figs. 5.1 and 5.2. On the other hand, first-order transitions emerge for high α_0, with discontinuity in the order parameters Q and R but continuous internal energy.

5.5 Final Remarks and Conclusions

The theoretical achievements of the seminal works [1–10] constitute a landmark in the study of multimode laser systems with disorder or not, in open or closed cavities.

Almost a decade after the first works reviewed in the present chapter, an RSB spin-glass phase in a photonic RL system was experimentally demonstrated by Ghofraniha and collaborators [22],

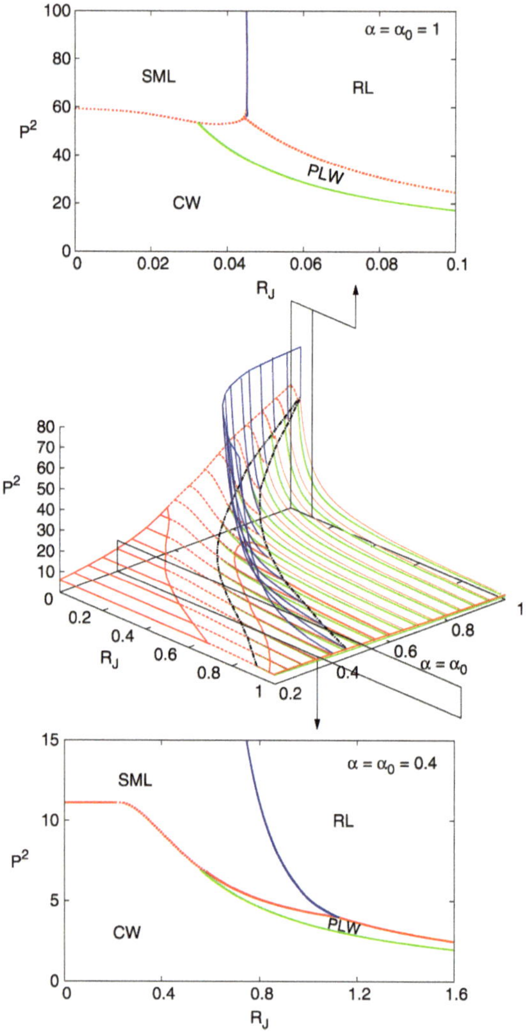

Figure 5.1 Phase diagrams in the ($\mathcal{P}^2$, R_J, $\alpha = \alpha_0$) space. For low disorder, only the SML and CW phases occur, varying pumping and degree of linearity. As disorder increases, the intermediate PLW phase arises between CW and SML. For strong disorder, RL replaces SML above the pumping threshold line. For any R_J, for low α_0 the transition driven by $\mathcal{P}$ is continuous in the order parameters, whereas for high nonlinearity ($\alpha > \alpha_{\text{nl}} = 0.6297$), it is discontinuous. Insets: phase diagrams in the ($\mathcal{P}^2$, R_J) plane for closed ($\alpha = \alpha_0 = 1$) and open ($\alpha = \alpha_0 = 0.4$) cavities. (Reproduced with permission from Ref. [5].)

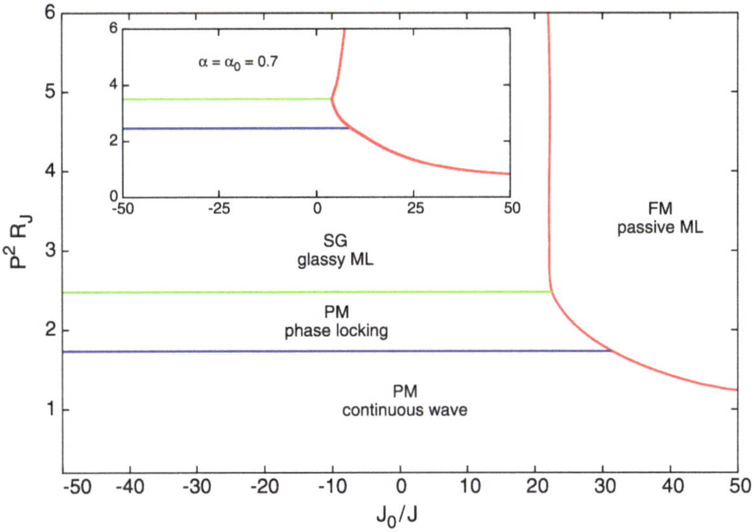

Figure 5.2 Phase diagram in $\mathcal{P}^2 R_J = \beta\epsilon^2 J$ vs $J_0/J = 1/R_J$ for closed cavity $\alpha = \alpha_0 = 1$ for $J_0 = J$ ranging from negative to positive. Inset: open cavity with $\alpha = 0.7$. (Reproduced with permission from Ref. [5].)

using a 2D functionalized T_5OC_x oligomer amorphous solid-state RL. In the subsequent years, a boost of works [23–37] appeared in the literature reporting on RSB glassy phases in many RL systems, with a diversity of structures and components, by our and other research groups (for a review on the work by Ghofraniha et al., see Chapter 6 of this book; also, Chapters 7 and 8 for further information on photonic RSB spin-glass phases in other materials). The number of articles published every year that characterize glassy behavior in actual systems keeps on increasing, thus consolidating the universality of the RSB spin-glass phase in photonic multimode laser systems with disorder, such as RLs and random fiber lasers.

Apart from the discovery and characterization of novel phases and photonic behaviors in RL systems, the pioneering works described in this chapter also opened up new avenues to explore RLs and random fiber lasers as platforms for the study of rather complex behaviors in quite different areas, such as nontrivial statistics, turbulence, magnetism, and complex systems. Therefore, they have served as inspiration for many research groups to advance the general understanding of RL systems.

Acknowledgments

We would like to thank the financial support from the Brazilian agencies Conselho Nacional de Desenvolvimento Científico e Tecnológico (CNPq), Coordenação de Aperfeiçoamento de Pessoal de Nível Superior (CAPES), and Fundação de Amparo à Ciência e Tecnologia do Estado de Pernambuco (FACEPE), through the project "Modelagem de Processos e Fenômenos Físicos em Materiais e Sistemas Complexos".

References

1. Angelani, L., Conti, C., Ruocco, G. and Zamponi, F. (2006). Glassy behavior of light. *Phys. Rev. Lett.*, **96**, 065702.
2. Angelani, L., Conti, C., Ruocco, G. and Zamponi, F. (2006). Glassy behavior of light in random lasers. *Phys. Rev. B*, **74**, 104207–104223.
3. Leuzzi, L., Conti, C., Folli, V., Angelani, L. and Ruocco, G. (2009). Phase diagram and complexity of mode-locked lasers: from order to disorder. *Phys. Rev. Lett.*, **102**, 083901.
4. Conti, C. and Leuzzi, L. (2011). Complexity of waves in nonlinear disordered media. *Phys. Rev. B*, **83**, 134204-134223.
5. Antenucci, F., Conti, C., Crisanti, A. and Leuzzi, L. (2015). General phase diagram of multimodal ordered and disordered lasers in closed and open cavities. *Phys. Rev. Lett.*, **114**, 043901.
6. Antenucci, F., Crisanti, A. and Leuzzi, L. (2015). Complex spherical 2 + 4 spin glass: a model for nonlinear optics in random media. *Phys. Rev. A*, **91**, 053816.
7. Antenucci, F., Crisanti, A., Ibañez-Berganza, M., Marruzzo, A. and Leuzzi, L. (2016). Statistical mechanics models for multimode lasers and random lasers. *Philos. Mag.*, **96**, 704-731.
8. Antenucci, F., Crisanti, A. and Leuzzi, L. (2015). The glassy random laser: replica symmetry breaking in the intensity fluctuations of emission spectra. *Sci. Rep.*, **5**, 16792.

9. Antenucci, F. I., Berganza and Leuzzi, L. (2015). Statistical physics of nonlinear wave interaction. *Phys. Rev. B*, **92**, 014204.
10. Tommasi, F., Ignesti, E., Lepri, S. and Cavalieri, S. (2016). Robustness of replica symmetry breaking phenomenology in random laser. *Sci. Rep.*, **6**, 37113.
11. Mézard, M., Parisi, G. and Virasoro, M. A. (1987) *Spin-Glass Theory and Beyond* (World Scientific, Singapore).
12. Parisi, G. (1979). Infinite number of order parameters for spin-glasses. *Phys. Rev. Lett.*, **43**, 1754–1756.
13. Parisi, G. (1980). A sequence of approximated solutions to the S-K model for spin glasses. *J. Phys. A*, **13**, L115–L121.
14. Hackenbroich, G., Viviescas, C. and Haake, F. (2002). Field quantization for chaotic resonators with overlapping modes. *Phys. Rev. Lett.*, **89**, 083902.
15. Viviescas, C. and Hackenbroich, G. (2003). Field quantization for open optical cavities. *Phys. Rev. A*, **67**, 013805.
16. Hackenbroich, G. (2007). Statistical theory of multimode random lasers. *J. Phys. A*, **38**, 10537.
17. Scully, M. and Zubairy, M. (1997) *Quantum Optics* (Cambridge University Press, Cambridge).
18. Jaynes, E. and Cummings, F. W. (1963). Comparison of quantum and semiclassical radiation theories with application to the beam maser. *Proc. IEEE*, **51**, 89–109.
19. Gardiner, C. W. and Zoller, P. (2010) *Quantum Noise* (Springer, Berlin).
20. Cugliandolo, L. F., Grempel, D. R., Lozano, G., Lozza, H. and da Silva Santos, C. A. (2002). Dissipative effects on quantum glassy systems. *Phys. Rev. B*, **66**, 014444.
21. Crisanti, A. and Sommers, H.-J. (1992). The spherical p-spin interaction spin glass model: the statics. *Z. Phys. B*, **87**, 341–354.
22. Ghofraniha, N., Viola, I., Di Maria, F., Barbarella, G., Gigli, G., Leuzzi, L. and Conti, C. (2015). Experimental evidence of replica symmetry breaking in random lasers. *Nat. Commun.*, **6**, 6058.
23. Lima, B. C., Gomes, A. S. L., Pincheira, P. I. R., Moura, A. L., Gagné, M., Raposo, E. P., de Araújo, C. B. and Kashyap, R. (2017). Observation of Lévy statistics in one-dimensional erbium-based random fiber laser. *J. Opt. Soc. Am. B*, **34**, 293-299.
24. De Araújo, C. B., Gomes, A. S. L. and Raposo, E. P. (2017). Lévy statistics and glassy behavior of light in random fiber lasers. *Appl. Sci.*, **7**, 644.

25. Gomes, A. S. L., Lima, B. C., Pincheira, P. I. R., Moura, A. L., Gagné, M., Kashyap, R., Raposo, E. P. and de Araújo, C. B. (2016). Glassy behavior in a one-dimensional continuous-wave erbium-doped random fiber laser. *Phys. Rev. A*, **94**, 011801(R).
26. Raposo, E. P., Gomes, A. S. L., Moura, A. L., Fewo, S. I., Pincheira, P. I. R., Lima, B. C., Carreño, S. J. M., Silva, A. F., Maia, L. J. Q., Jerez, V., Gagné, M., Kashyap, R. and de Araújo, C. B. (2016). Universal phase transitions in random lasers. *Opt. Photon. News*, **27** (12), 34.
27. Gomes, A. S. L., Raposo, E. P., Moura, A. L., Fewo, S. I., Pincheira, P. I. R., Jerez, V., Maia, L. J. Q. and de Araújo, C. B. (2016). Observation of Lévy distribution and replica symmetry breaking in random lasers from a single set of measurements. *Sci. Rep.*, **6**, 27987.
28. Pincheira, P. I. R., Silva, A. F., Carreño, S. J. M., Moura, A. L., Fewo, S. I., Raposo, E. P., Gomes, A. S. L. and de Araújo, C. B. (2016). Observation of photonic paramagnetic to spin-glass transition in specially-designed TiO2 particles-based dye-colloidal random laser. *Opt. Lett.*, **41**, 3459–3462.
29. Moura, A. L., Pincheira, P. I. R., Reyna, A. S., Raposo, E. P., Gomes, A. S. L. and de Araújo, C. B. (2017). Replica symmetry breaking in the photonic ferromagneticlike spontaneous mode-locking phase of a multimode Nd:YAG laser. *Phys. Rev. Lett.*, **119**, 163902.
30. González, I. R. R., Raposo, E. P., Macêdo, A. M. S., Menezes, L. de S. and Gomes, A. S. L. (2018). Coexistence of turbulence-like and glassy behaviours in a photonic system. *Sci. Rep.*, **8**, 17046.
31. Raposo, E. P., González, I. R. R., Macêdo, A. M. S., Lima, B. C., Kashyap, R., Menezes, L. de S. and Gomes, A. S. L. (2019). Evidence of Floquet phase in a photonic system. *Phys. Rev. Lett.*, **122**, 143903.
32. Basak, S., Blanco, A. and López, C. (2016). Large fluctuations at the lasing threshold of solid- and liquid-state dye lasers. *Sci. Rep.*, **6**, 32134.
33. Pierangeli, D., Tavani, A., Di Mei, F., Agranat, A. J., Conti, C. and DelRe, E. (2017). Observation of replica symmetry breaking in disordered nonlinear wave propagation. *Nat. Commun.*, **8**, 1501.
34. Choubey, P. S., Sarkar, A., Varshney, S. K. and B. N., Shivakiran Bhaktha. (2020). Random laser spectroscopy and replica symmetry breaking phase transitions in a solvent-rich polymer thin film waveguide. *J. Opt. Soc. Am. B*, **37**, 2505–2513.
35. Sarkar, A., Andreasen, J. and Bhaktha, B. B. N. S. (2020). Replica symmetry breaking in a weakly scattering optofluidic random laser. *Sci. Rep.*, **10**, 2628.

36. Zhou, Z., Chen, L. and Bao, X. (2021). High efficiency Brillouin random fiber laser with replica symmetry breaking enabled by random fiber grating. *Opt. Express*, **29**, 6532–6541.
37. Kong, J., He, J., Zhang, J., Ma, J., Xie, K., Chen, J., Hong, L. and Hu, Z. (2021). Replica symmetry breaking in cholesteric liquid crystal bandgap lasing. *Ann. Phys.*, **533**, 2000328.

Chapter 6

Spin Glass Behavior of Random Lasers: Experiments

Neda Ghofraniha[a] and Claudio Conti[b]

[a]*Istituto dei Sistemi Complessi (CNR-ISC)*
c/o Dipartimento di Fisica-Università La Sapienza
Piazzale A, Moro 2, 00185-Roma, Italia
[b]*Dipartimento di Fisica-Università La Sapienza*
Piazzale A, Moro 2, 00185-Roma, Italia
neda.ghofraniha@cnr.it

6.1 Introduction

In 2015, Ghofraniha et al (Ghofraniha et al., 2015) reported the first experimental evidence of replica symmetry breaking (RSB) in a random laser (RL). In the last four years, many different authors have shown different aspects of this intriguing topic in diverse RLs. In this chapter, we summarize these findings after a brief and simple description of RSB and spin glass theory, aimed toward experimentalists and researchers not specialized in statistical mechanics theory.

Lévy Statistics and Spin Glass Behavior in Random Lasers
Edited by Anderson S. L. Gomes, André L. Moura, Cid B. de Araújo, and Ernesto P. Raposo

ISBN 978-981-4968-55-3 (Hardcover), 978-1-003-33618-1 (eBook)
www.jennystanford.com

6.2 Spin Glass and Replica Symmetry Breaking

Spin glasses (Castellani and Cavagna, 2005; Mézard et al., 1987) are disordered magnetic systems in which two nearby localized magnetic moments interact either ferromagnetically (with aligned moments) or antiferromagnetically (with antialigned moments).

The term "glass" comes from the analogy between the magnetic disorder in a spin glass and the positional disorder in a conventional, molecular or colloidal glass. In window glass or any amorphous solid, the atomic bond structure is highly irregular; in contrast, a crystal has a periodic pattern of atomic bonds. Similarly, in a ferromagnetic solid magnetic spins are all aligned.

Spin glasses belong to the class of disordered systems with "quenched disorder." In these systems the disorder is explicitly present in the Hamiltonian H, under the form of random couplings J among spin moments σ, as $H = H(\sigma, J)$. A famous example is the Edwards-Anderson model [5],

$$H = -\sum_{<ij>} J_{ij}\sigma_i\sigma_j, \qquad (6.1)$$

where the couplings J_{ij} are Gaussian random variables. Equation (6.1) is a finite dimensional model, since the sum is performed over nearest-neighbor spins. The disorder is quenched, when the couplings J_i are constant on the time scale over which the spins σ_i fluctuate.

Disordered patterns of aligned and misaligned atomic magnets in spin glasses create frustration: For each spin, it becomes impossible to satisfy all the couplings at the same time, as in a ferromagnetic system. Figure 6.1 shows three spins on a triangle with two positive and one negative coupling. No single configuration of spin alignment will satisfy all the interactions, minimizing the Hamiltonian. The system will settle in one of the two degenerate configurations with the lowest realizable energy.

Frustration creates many competing arrangements of atoms. This is the main reason for the proliferation of metastable states in disordered systems. "Metastable" means that such systems are "stuck" in many stable and different configurations other than the ferromagnetic lowest-energy configuration. For this reason, at low

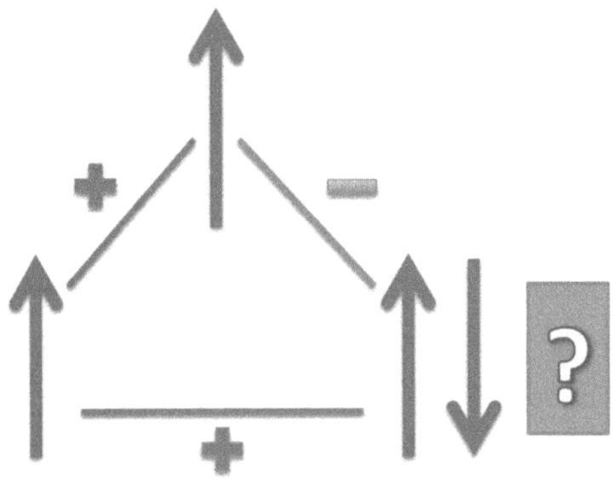

Figure 6.1 Scheme of the frustration between spins.

temperature ergodicity breaking occurs: The system at equilibrium explores only a sub-part of the phase space, constituted by the so-called pure states. We can image the phase space as decomposed in valleys (phases, states) separated by barriers. If the free energy as a function of the configuration space has many minima (a corrugated free energy landscape), the number of the states is very large. In mean-field systems (infinite dimension), barriers between states are infinite in the thermodynamic limit. In contrast, systems with finite-dimensions cannot remain trapped for an infinite time in a metastable state, because free energy barriers are finite. Therefore, the system, initially trapped in a metastable state, will, sooner or later, collapse in the stable state with lower free energy.

How do we deal with the disorder? The problem is that any observable depends on the particular realization of the disorder J, including the free energy of the system. This is analytically solved by the **replica method**, where the relevant thermodynamic quantities are evaluated by considering n identical replicas (i.e., copies) of the system that explores the many states of the phase space. In the replica method, any average is done by using the following trick:

$$\overline{\log Z^n} = \lim_{n \to 0} \frac{1}{n} \log \overline{Z^n} \tag{6.2}$$

with Z the partition function.

In case of ergodicity breaking and many pure thermodynamic states, it is possible to assign each configuration in the phase space with nonzero thermodynamic (Gibbs) weight to one and only one state. In this case we can write for any observable A, the average $< A > = \sum_{\alpha} < A >_{\alpha} w_{\alpha}$, with $w_{\alpha} = Z_{\alpha}/Z$ the statistical weight of state α. In a finite dimension, the decomposition in pure states is only valid for finite times, i.e., times much smaller than the time needed for the stable equilibrium state to take over. On the other hand, in mean-field systems (infinite dimension), barriers between metastable states may be infinite in the thermodynamic limit, and the metastable states can be considered pure states with Gibbs weight w_{α}.

To study phase transitions in glasses, a measure of the similarity between the accessible states is needed, this is called "overlap" and the general form is:

$$q_{\alpha\beta} = \sum_{i=1}^{N} \sigma_i^{\alpha} \sigma_i^{\beta}, \tag{6.3}$$

where the sum is over the N configurations (spins) in the states α and β (Parisi, 1980, 1983). Another powerful trick is considering many physical systems with the same quenched disorder, or replicas. α and β become replica indexes and the overlap among states translates into the overlap among replicas. In mean-field spin-glasses, there are many non-equivalent pure states at low temperatures. In this case, it is useful to introduce the probability distribution of all the possible values of the overlaps among replicas, defined as:

$$P(q) = \sum_{\alpha\beta} w_{\alpha} w_{\beta} \delta(q - q_{\alpha\beta}), \tag{6.4}$$

where w_{α} and w_{β} are the weighs of the replicas and $\delta(q - q_{\alpha\beta})$ is the delta function peaked at $q_{\alpha\beta}$.

Spin glass theory predicts that the statistical distribution $P(q)$ of the order parameter in Eq. (6.3), namely the "Parisi overlap," changes shape when a large number of competing equilibrium states emerges in the energetic landscape. When this happens, "replicas of the system," i.e., identical copies under the same experimental conditions, may furnish different values of their mutual correlations (overlaps), depending on whether or not their equilibrium dynamics thermalizes in the same ergodic components or in separate ones.

In essence, if there is more than one state, $P(q)$ is not a single delta function and we say that the replica symmetry is broken: Two identical replicas of the same system may stay in a quite different state.

Spin glasses can be achieved in laboratories in several ways. The most common is to substitute, at random locations, a very small concentration (at most a few percent) of a magnetic element, often iron or manganese, inside a nonmagnetic metallic host, such as one of the noble metals (copper, silver, or gold). Depending on the placement of two nearby magnetic impurities, the conduction-mediated magnetic interaction can therefore be either ferromagnetic or antiferromagnetic. Unfortunately none of the realized systems confirms the main spin glass theoretical paradigm that is the replica symmetry breaking (RSB).

Why are spin glasses interesting? There are several reasons. From the point of view of physics, spin glasses and other systems with quenched disorder represent a serious gap in our understanding of condensed matter. In more conventional systems, long-range order and useful symmetries enable us to use many of the powerful tools of condensed matter physics and statistical mechanics to establish a conceptual framework to understand their nature and behavior. Such notions include Bloch waves, broken symmetry, order parameters, Goldstone modes, topological singularities in order-parameter space, and many others. The study of spin glasses has led to a surprising variety of applications to problems in biology, computer science, economics, and very recently in random photonics. The usefulness of spin glass concepts, bridging to fields outside of physics, is one of the early reasons spin glasses came to be regarded as relevant to the study of complexity.

6.3 Random Lasers and Spectral Fluctuations

Random Lasers (RLs) are cavityless lasers where a gain material is dispersed in a disordered system that acts as a resonator (see the scheme in Fig. 6.2). Under an external pumping (e.g., a laser) emitted light can be partially entrapped thanks to multiple scattering leading to stimulated emission and thus lasing. In RLs,

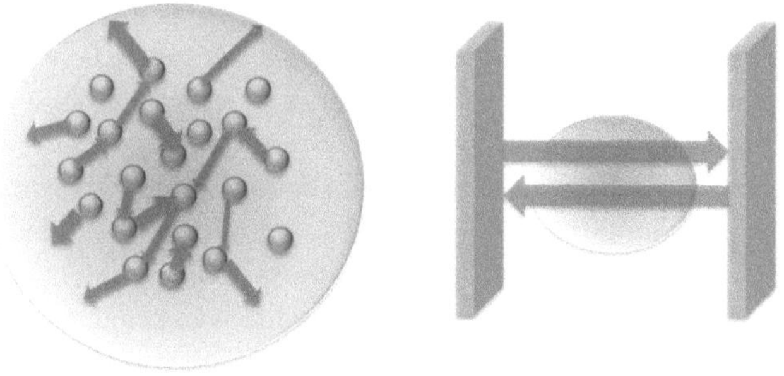

Figure 6.2 Schematics of feedback mechanism in standard (right) and random (left) lasers.

there is no a rigid cavity with well-defined electromagnetic (e.m.) modes like in standard lasers and light follows different paths each time the gain is switched on, leading to the emission of various combinations of activated modes. Such random behavior recalls some specific characteristics of complex glassy systems and more than 10 years ago the glassy behavior of light in RLs has been introduced theoretically (Angelani et al., 2006).

A manifestation of that is the spectral fluctuations observed in some RLs and this is the subject of this book chapter.

Shot-to-shot fluctuations in RLs have been reported by various authors in the last decade. In the study by Fallert et al. (Fallert et al., 2009), they are ascribed to the coexistence of localized and extended modes without analyzing the statistical distribution of their shot-to-shot correlations. Moreover, the fluctuations have been described by means of mode repulsion (Cao et al., 2003) and in Molen et al.'s study (van der Molen et al., 2006), they are ascribed to the intrinsic matter fluctuations in fluids. Levy statistics (Lepri, 2013; Lepri et al., 2007; Mujumdar et al., 2007; Raposo and Gomes, 2015; Uppu and Mujumdar, 2010) has been also extensively used to explain RL spectral fluctuations. In Mujumdar et al.'s study (Mujumdar et al., 2007), mode-competition and seemingly chaotic behavior have been considered, and intended as a generic sensitivity to initial conditions. In a paper by Skipetrov et al. (Skipetrov and Maynard, 2000), fluctuations due to a transparent Kerr medium are theoretically considered, and even if they are not strictly related to the case with gain and loss of RLs, they look to have connections with the spin-glass approach in the study by Leuzzi et al. (Antenucci et al., 2015a; Conti and Leuzzi, 2011; Leuzzi et al., 2009). The latter approach has been developed in a series of papers, and is based on the mode-coupling laser theory as, for example, in the study by Leonetti et al. (Leonetti et al., 2013), which reports on the time dynamics of frequency locking mechanism but without addressing the statistical properties of the laser emission.

In the recent years, the first explanation has been given by the Replica Symmetry Breaking theory (Antenucci et al., 2015b,c; Ghofraniha et al., 2015): From a statistical mechanical point of view, the RL spectral fluctuations are a manifestation of the existence of many degenerate lasing states, each one corresponding to a given ensemble of activated modes with their own wavelengths, phases and intensities. This theory is obtained starting from the decomposition in normal modes of the electromagnetic field of light, in the presence of non-linearity, and leading to a quenched disordered spin-glass-like Hamiltonian (Antenucci et al., 2015c). The theoretical approach considers the combined effect of localized and radiating modes as occurring in any open cavity and allows to describe open systems with arbitrary degree of openness (Antenucci et al., 2015a). A complex landscape typical of glassy systems is expected to occur beyond some threshold critical value of the external parameter driving the transition. The measurements reported in (Ghofraniha et al., 2015) confirmed for the first time this theory.

6.4 First Experimental Prove of RSB in RLs

In the paper (Ghofraniha et al., 2015), the authors use a thiophene (T5OCx) dye chemically treated in order to have a tick disordered solid material (Fig 6.3a) that shows evident and strong shot to shot fluctuations in the emission intensity.

A sketch of the pumping and collecting geometry is given in the inset of Fig. 6.3b.

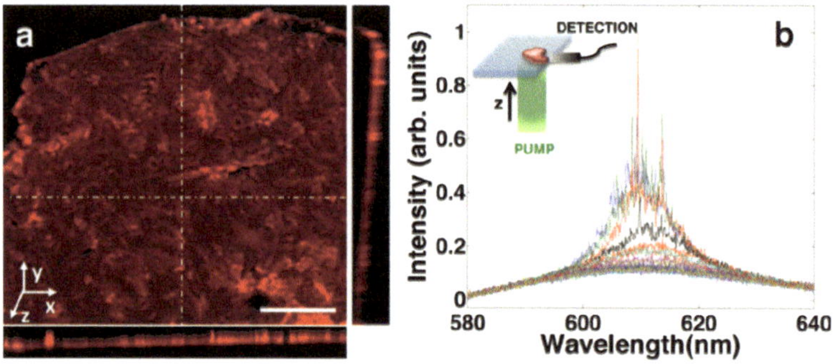

Figure 6.3 Sample image and emission spectra showing random lasing. (a) 3D-reconstruction of confocal microscopy Z-stack images of the supramolecular laminar packing in aT5OCx solid sample. The right and the bottom panels report the yz- and the xz-sections, respectively. Scale bar: 20 μm. (b) High-resolution single-shot spectra taken in the same conditions, 10 mJ pump energy. Inset: sketch of the experiment. Reproduced from Ref. (Ghofraniha et al., 2015) with permission from Nature Publishing Group.

In Fig. 6.3b, single-shot emission spectra with high resolution (0.07 nm), taken at identical experimental conditions, are illustrated. The input energy is 10 mJ. The presence of RL modes with configuration variable from pulse to pulse is evident: Each time the system is pumped, the numerous passive modes randomly compete for the available gain, giving rise to several different compositions of the activated spectral peaks (Mujumdar et al., 2007). Such behavior is also evidenced by the direct visualization of the sample during pumping, as reported in Fig. 6.4, where four different fluorescence images taken at four single shots exhibit different emission patterns and corresponding spectra. The spatial intensity profile of the lasing modes changes from shot to shot and many extended modes are concentrated in different extended spots. From shot to shot the size and the brightness of these spots change showing that also the spatial structure of the modes changes.

In Figs. 6.5a and 6.5b emissions of subsequent 100 shots with lower spectral resolution (0.3 nm) and at two different pump energies are shown. The spectra fluctuate randomly from pulse to pulse and at low energy the noisy variations of the spontaneous

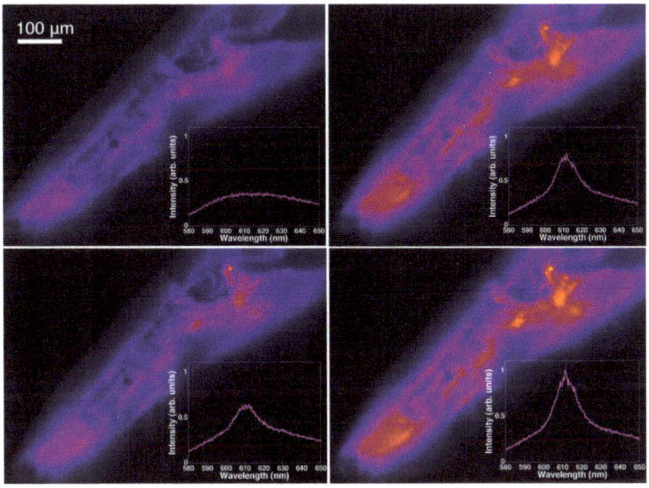

Figure 6.4 Snapshots of RL emissions. Single-shot optical images and corresponding emission spectra (insets) during the pumping of the sample in the same experimental conditions. The input energy is 10 mJ. Reproduced from Ref. (Ghofraniha et al., 2015) with permission from Nature Publishing Group.

emission are negligible if compared to the pronounced fluctuations observed at high energy.

The experimental results are analyzed for the first time by means of the replica theory. In this framework the cavity coupled modes in the Hamiltonian describing random lasers (Leuzzi et al., 2009) are variables not easily accessible in the experiments, but from the spectra it is possible to have access to the intensities $I(k) \propto \langle |a_j^2| \rangle_k$, where a_j is the amplitude of the longitudinal mode j (Conti and Leuzzi, 2011; Leuzzi et al., 2009) at wavelength λ_j. In this analysis, each spectrum represents a different state of the same thermodynamic phase. The average is taken over all modes having similar wavelength, binned together in data acquisition with index k, due to the finite spectral resolution $\Delta\lambda = 0.3$ nm. The average spectral intensity is defined as

$$\epsilon_\alpha = \frac{1}{N} \sum_{k=1}^{N} I_\alpha(k), \tag{6.5}$$

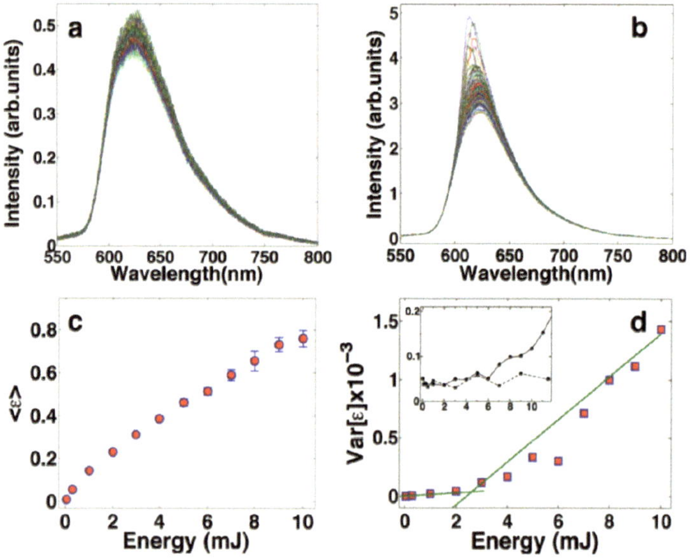

Figure 6.5 Fluctuations in the random laser emission. (a, b) Emission spectra at low energy 1 mJ (a) and high energy 12 mJ (b). (c) Mean emission intensity $<\epsilon>$ vs. pump energy, error bars are standard deviation. (d) Variance of ϵ vs. pump energy. Inset: peak standard deviation divided by mean peak intensity from the sample (solid line) and the pump laser (dashed line) emission spectra. Reproduced from Ref. (Ghofraniha et al., 2015) with permission from Nature Publishing Group.

being α the replica index and N the number of acquired spectral points. The variation of this parameter from shot-to-shot quantifies the observed fluctuations. Mean and variance of ϵ_α over N_s replicas for each pumping are

$$<\epsilon>=\frac{1}{N_s}\sum_{\alpha=1}^{N_s}\epsilon_\alpha, \tag{6.6}$$

$$\mathrm{Var}[\epsilon]=\frac{1}{\mathrm{N_s}}\sum_{\alpha=1}^{\mathrm{N_s}}(\epsilon_\alpha-<\epsilon_\alpha>)^2. \tag{6.7}$$

Figure 6.5c shows the continuous growth for increasing input energy of the mean intensity, Eq.(6.6), that is signature of amplified spontaneous emission (ASE) (Anni et al., 2004; Frolov et al., 1999), characterized by the absence of threshold in the mean spectral intensity $< \epsilon >$. In addition to ASE, RL is evident in the spectra from the fine peaks superimposed to the ASE band in Fig. 6.4b. It is well known that ASE does not exhibit a threshold, at variance with RL and in this case the RL threshold in the emission intensities is masked by ASE when coming at the spectral peak. The RL threshold was, in this system of thick amorphous solid dye, identified from the spectral fluctuation, quantified by Var[ϵ], calculated with Eq. (6.7) and shown in Fig. 6.5d. Such behavior has not been reported in previous works on similar dyes (Anni et al., 2004; Frolov et al., 1999). The threshold of the fluctuations in Fig. 6.5d reveals the onset of the RL action, that for this sample is estimated around 3 mJ.

All experiments are performed in the same environmental conditions, that is the temperature, the position and the state of the sample and the configuration of the optical components are controlled to be stable by growing pump energy. To check the stability of the input laser, the authors estimate the energy fluctuations as the percentage variation given by the ratio between the standard deviation and the mean value of the emission peak intensity with and without sample, for increasing energy. The plots are reported in the inset of Fig. 6.5d, where the laser energy variations are approximately around 4% at any energy, while the same quantity from the sample increases at high energy. We stress that an error of 4% on the x-axis of Fig. 6.5d is inside the symbols and negligible compared to the energy values, thus it cannot cause the evident increase of the observed variance.

To analyze the system behavior in the framework of statistical mechanics of disordered systems, and characterize the high pumping RL regime, an order parameter to identify the phase transition is introduced: the overlap between intensity fluctuations in different experimental replicas. The experimentally accessible variable, coarse graining the behavior of single modes, is the intensity fluctuation at a given frequency:

$$\Delta_\alpha(k) = I_\alpha(k) - \bar{I}(k) \tag{6.8}$$

with $\bar{I}(k)$ the average over replicas of each mode intensity

$$\bar{I}(k) = \frac{1}{N_s} \sum_{\alpha=1}^{N_s} I_\alpha(k). \tag{6.9}$$

The overlap between pulse-to-pulse intensity fluctuations is defined as

$$q_{\alpha\beta} = \frac{\sum_{k=1}^{N} \Delta_\alpha(k)\Delta_\beta(k)}{\sqrt{\sum_{k=1}^{N} \Delta_\alpha^2(k)}\sqrt{\sum_{k=1}^{N} \Delta_\beta^2(k)}}. \tag{6.10}$$

From the measured spectra the set of all $N_s(N_s - 1)/2$ values q of $q_{\alpha\beta}$ for each different input energy are calculated, determining their distribution $P(q)$.

$P(q)$ is considered as a measure of the theoretical distribution of the overlap between mode amplitudes (Conti and Leuzzi, 2011; Leuzzi et al., 2009), which is a leading quantity for the description of glassy phases (Mézard et al., 1987). No experimental measurements of $P(q)$ have been reported before (Ghofraniha et al., 2015).

Six examples are reported in Fig. 6.6a–f for increasing pump energy. At low energy (Fig. 6.6a), all overlaps are centered around the zero value, meaning that the electromagnetic modes are independent and not interacting in the paramagnetic regime. By increasing energy, modes are coupled by the nonlinearity and this corresponds to a non-trivial overlap distribution (Fig. 6.6b–f). In the high-energy glassy phase, with all modes highly interacting and frustrated by the disorder, q assumes all possible values in the range [−1,1]. Such behavior of the $P(q)$ evidences the fact that the correlation between intensity fluctuations in any two replicas depends on the replicas selected. The variety of possible correlations extends to the whole range of values. This is a manifestation of the breaking of the replica symmetry.

In Fig. 6.6g another useful parameter $q_{\max}$ corresponding to the position of the maximum of $P(|q|)$ versus pumping is shown: It changes drastically signaling a phase transition between 2 mJ and 3 mJ, compatible with the threshold determined by Var[ϵ] in Fig. 6.6d. We point up that in the high pump regime $P(q) > 0$ for any value of q.

The parameter q measures the overlap of the shot-to-shot fluctuations at each wavelength around the average spectrum at a given pumping. Thus $q \sim 1$ when two shots both have very similar intensity fluctuations at each wavelength. In principle, it does not matter how large the fluctuations are to have a large q, but only how much they are correlated. The fact that large overlap

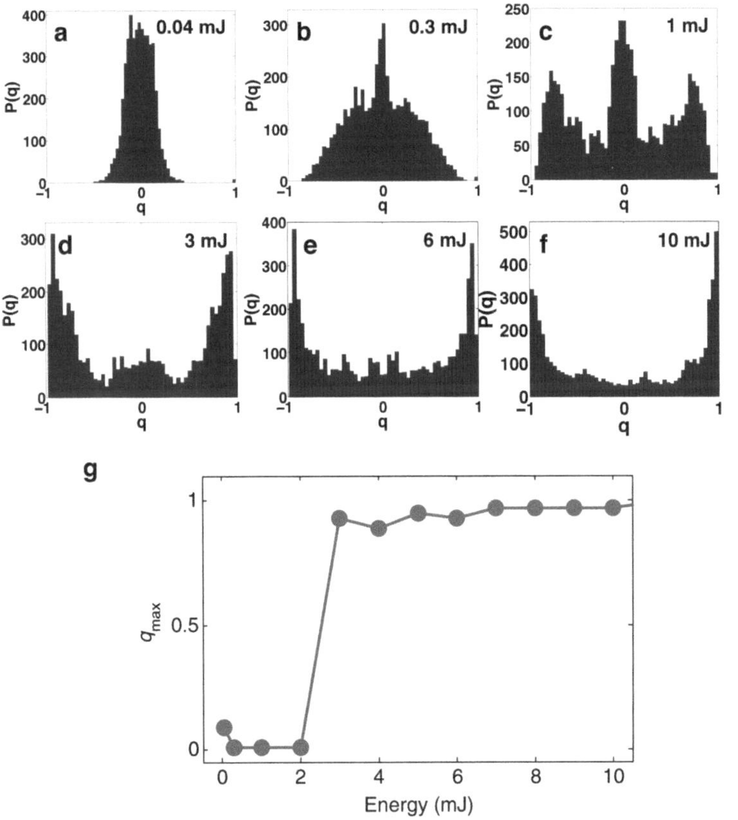

Figure 6.6 (a-f) Distribution function of the overlap showing replica symmetry breaking by increasing pump energy. (g) $q_{\max}$ corresponding to the position of the maximum of $P(|q|)$ versus pumping. Reproduced from Ref. (Ghofraniha et al., 2015) with permission from Nature Publishing Group.

values become probable in the lasing regime above threshold is not a trivial feature. Similarly, one has $q \sim -1$ when two shots display spectral fluctuations similar in magnitude (no matter how large) but opposite in sign, that is the shot-to-shot fluctuations around the average spectrum are anti-correlated. When $q \sim 0$ the intensity fluctuations in two shots are uncorrelated. This can be due to the ASE background, but can also correspond to the presence of a sub-set of shots with uncorrelated large fluctuations. Above threshold, the latter situation occurs, as can be observed looking at the single spectra. By the definition of overlap in Eq. (6.10) $|q| \sim [0, 1]$, the fluctuations of a single spectrum can be large or small. The important observation is that increasing the pumping, the correlations between shot-to-shot fluctuations become non-trivial right when the fluctuation magnitude starts increasing significantly (such as above threshold), see Fig. 6.5d.

To stress the peculiarity of such a signature of RSB and the relevance of the adopted method in probing the onset of a glassy nature of RLs, the $P(q)$ for the emission spectra of a standard ordered laser (the one used as pump laser in the experiments) are calculated and reported in Fig. 6.7a. The same is done for a differently structured RL, a liquid dispersion of titanium dioxide in 3 mM Rhodamine B-ethylene glycol solution, as shown in Fig. 6.7c,e,g. In Fig. 6.7b the standard laser (obviously well above threshold) does not show any symmetry breaking, even though small fluctuations are present, as expected in an ordered system. In Fig. 6.7d,f,h the $P(q)$ for the liquid RL are displayed by increasing pump energy across its estimated threshold of 0.25 mJ and no evident signs of replica symmetry breaking above the lasing threshold can be appreciated up to 0.85 mJ. This result points out that indications about the occurrence of RSB depend on the characteristics of the system and are, therefore, a way to classify the nature of different RLs.

More generally, RSB can occur in any RL where intensity fluctuations grow drastically as the pump energy increases beyond a threshold, exhibiting a transition from negligible to large oscillations. A possible order parameter for such transition is quantified by the variance of the fluctuations and a more refined one is the distribution of all mutual overlaps between intensity fluctuations in different physical replicas. Besides evidencing the threshold of phase transition, it describes the organization of states and their non-trivial correlation, non-symmetric in replicas exchange. Above the threshold pump energy, the overlap between two replicas does depend on the specific couple chosen and the variation from couple to couple turns out to be more complicated than the ordinary statistical fluctuations. That is, replicas are not all equivalent to each other in the analyzed random laser and , thus, an instance of

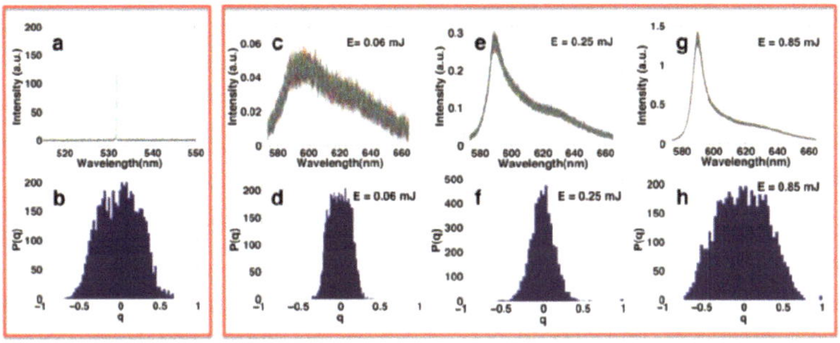

Figure 6.7 Distribution function of the overlap in standard and random laser without replica symmetry breaking. (a, b) Emission spectra (a) and $P(q)$ (b) of a Q-switched pulsed Nd-YAG standard ordered laser. The analysis is done on 100 shots. (c–h) Emission spectra (c, e, g) and correspondent $P(q)$ (d, f, h) of a liquid dispersion of titanium dioxide in Rhodamine B-ethylene glycol solution at three different pump energy through the threshold.

replica symmetry breaking through the random laser transition can be observed.

In the pioneering work (Ghofraniha et al., 2015) not only a first experimental proof of RSB was presented, but a new method is provided that allows to characterize the nature of the randomness above the lasing threshold applicable to the whole profusion of random lasers. By directly probing the measure of spectral intensity fluctuations, the possible existence and the organization of many equivalent competing lasing states yielding a glassy light behavior can be verified, allowing for an alternative, supplementary RL classification in terms of glassiness.

It is important to recall that in the effective statistical mechanical Hamiltonian for RLs the variables are, actually, the modes, such as complex amplitudes (Antenucci et al., 2015a; Conti and Leuzzi, 2011; Leuzzi et al., 2009), not easily accessible in the experiments. Whereas their squared modulus given by the emitted intensity are considered, thus losing information about the actual state structure. In a recent work (Antenucci et al., 2015c), it is demonstrated that such coarse-graining appears to be refined enough to investigate the peculiar properties of glassy-like regimes of random lasers.

6.5 Evidence of RSB in Other Different RLs

Before reporting the main results of RSB in other kinds of RL, it is important to stress the concept that not all random lasers show strong emitted intensity fluctuations and consequent breaking of replica symmetry. In (Ghofraniha et al., 2015), it has been demonstrated that a dispersion of titanium dioxide (TiO_2) in a solution of Rhodamine B under high pumping shows RL behavior but with very slight spectral fluctuations, in this system the similarity (symmetry) between the numerous taken spectra (replica) is strong and conserved. In this work, the authors claim that such observation is due to the structure of the material that acts as laser cavity. The solid thiophene one shows RSB because of the presence of many interacting modes while the fluid sample has lower light scattering ability and thus is a less efficient cavity with many modes presumably not in strong competition. Such classification is not strict, the relevant point is not the physical phase of the RL material but rather its microscopic structure, morphology and refractive capacity. It is hard to do a sharp division between the several realized RL systems and it is more correct to state that RSB is characteristics of those RLs where the scattering strength, quantified by the mean free path of the light traveling in the sample, is in an optimal range to have many electromagnetic modes (not single modes) and in strong interaction mediated by the gain medium. Theoretical phase diagrams have been proposed in past (Leuzzi et al., 2009), but to our knowledge there is not a diagram based on experiments.

Within this discussion, RSB has been evidenced in liquid phase samples, for instance, in (Pincheira et al., 2016) and (Tommasi et al., 2016). In (Pincheira et al., 2016), specially designed amorphous TiO_2 particles are synthesized by a sol-gel method and suspended in an ethanol-Rhodamine 6G solution. The particles are then chemically functionalized to minimize precipitation in the timescale of the experiments and the so obtained fluid material is put in cuvette glass and impinged by the second harmonic at 532 nm of a Nd-YAG laser operating at 5 Hz and 7 ns pulse width. The dyed colloidal system shows clear incoherent random laser behavior. The data are then analyzed by following the same procedure in (Ghofraniha et al., 2015) based on the calculation of the Parisi overlap defined by Eq. (6.10) and RSB is evidenced by the results. In this work the authors do not deepen the fact that in liquid materials the structural composition changes in time due to the Brownian motion of the particles and the system is not under the same conditions during the measurements, thus single spectra from single pump shots cannot be properly considered as replicas.

This crucial point is investigated in Ref. (Tommasi et al., 2016), where the random laser consists of a solution of dye in ethanol with ZnO nanoparticles as light scatterers. In this paper the authors analyze the speckle patterns from the light scattered by the disordered material and show speckle movements during the total time of acquisition and from shot to shot. They conclude that the change in the speckle patterns between two consecutive shots is signature of the lack of the condition of quenched disorder for the propagating radiation and that further theoretical effort is needed to find an explanation of the fact that RSB is observed also in a fluid random material with variable non-quenched disorder.

Our point of view is that in RLs with incoherent feedback characterized by broad spectral emission line also above threshold many longitudinal modes are present that cannot be distinguished due to the low instrumental resolution. These many modes, if in strong interaction, may compete for the available energy given by the gain and lead to the strong intensity fluctuations that grow with the pumping laser power and after some threshold become completely de-correlated and lose their self similarity. If, in addition, the configuration of the disorder in the material changes in time like in liquid samples further intensity fluctuation might be expected that should not lead to any replica symmetry breaking because in this case each single-shot spectrum cannot be considered as a replica.

However, also in such conditions in each different configuration, characterized by different speckle pattern, many and many modes are activated and once in strong coupling, they can give to intensity fluctuations much stronger than those due to the Brownian motion of the particles evidenced by speckle fluctuations. One more important point is that speckle patterns can be considered as indicators of space distribution of the passive transverse modes, that can vary from shot to shot due to the Brownian motion of the scattering particles. This can affect the distribution of the longitudinal modes, but not the entity of their interaction mediated by the gain and their average number, that remains very high.

In other words, for a fixed pumping energy, what changes from shot to shot in fluid samples is mainly the propagation path of light that can influence the number of the excited modes and their distribution both in time and in space. If the total number of the activated resonances is very high as in incoherent RLs such variations can be considered as small perturbations that do not affect the robustness of the replica symmetry breaking theory applied to disordered lasers. The extent of the fluctuations and their mutual correlations depend strongly on the energy given to the system, correspondent to the inverse temperature in spin glasses, and only marginally on the mutual position of the spins if we are in presence of a huge number of coupled spins. In conclusion, to our opinion in the mentioned examples and similar ones RSB theory remains robust and a glassy state of RL is still distinguished by it, because the coupling between modes (spins) is stronger than the variation of disorder between configurations that in this case can be considered and named quasi-replicas of the thermodynamic state. Experimentally the influence of disorder variation can be quantified by measuring and comparing the variance of the emission intensity fluctuations as in Fig. 6.5d above threshold of a RL with frozen and unfrozen disorder. Another test would be to estimate image correlations between shot to shot speckle patterns at variance with the input energy. To our knowledge, these investigations have not been yet done.

A different class of RLs showing RSB is furnished by erbium-based random fiber lasers (RFL) and widely investigated by A. Gomes and coworkers. A first demonstration in given in Ref. (Gomes et al., 2016a), where the authors employ an erbium-doped RFL (Er-RFL) in which scattering occurs from fiber Bragg Gratings (FBGs) randomly spaced and prepared in a unique way such as to provide a suitable density of scatterers (de Matos et al., 2007). Specifically, to increase the scattering strength of the material a polarization maintaining Er-doped fiber is used in which a continuous grating with randomly distributed phase errors is written, instead of a

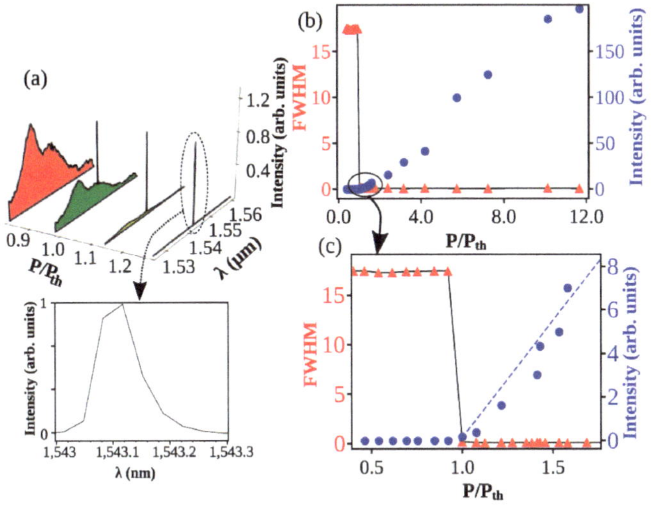

Figure 6.8 (a) Spectral evolution as a function of the pump power P, normalized by the threshold power P_{th}. (b) FWHM (triangles) and emitted intensity (circles) as a function of P/P_{th}. (c) Closer look at the threshold region in (b), showing the threshold inferred from the FWHM or output versus input intensity. Lines are guides to the eye. Reproduced from Ref. (Gomes et al., 2016a) with permission from American Physical Society.

random array of gratings. In the described experiments a 30 cm-long Er-RFL is used, with a continuous wave pump source operating at 1480 nm delivering up to 200 mW power. Measurements at 980 nm pump wavelength are also performed and provide similar results. The spectral characteristics of RL emission are reported in Fig. 6.8, where the sharp line broadening and intensity growth evidence random lasing with a defined threshold pump power.

Evidence of the stimulated emission in such system is given by a sharp resonance line observed in the emission spectra at high power that seems to be a single emission mode. For the statistical mechanical analysis it is important to stress (as the authors discuss and show in the paper) that this kind of RFL ,even though characterized by one sub-nanometric spectral line, is still a multimode laser, where the many modes are not observable due to the low spectral resolution. In order to confirm the multimode character of the Er-RFL in this work, the speckle contrast from far field transmission is measured by a CCD camera operating in the visible (400–900 nm) and in the near infrared (0.4–1.9 nm) and the number of modes is estimated as the ratio of mean and standard deviation of the speckle intensity. With this procedure they nicely show that standard pulsed Nd:YAG and semiconductor lasers have few distinct (2–3) modes and their one dimensional Erbium random fiber laser has more than 200 longitudinal modes, similarly to well-known RLs as a dispersion of TiO_2 in Rhodamine 6G dye solution. The emission intensity fluctuation is analyzed after 1500 acquisition spectra (replica) for different pumping power and the Parisi overlap q is measured from Eq. (6.10). In Fig. 6.9, the overlap distribution $P(q)$ is depicted together with the parameter q_{max}, at which $P(q)$ assumes the maximum, versus normalized pump power in comparison with the FWHM of the spectral lines. It is demonstrated that both sharp line narrowing and q_{max} growth from 0 to 1 occur at the same identical input power that is the laser threshold. This confirms the findings in Ref. (Ghofraniha et al., 2015).

6.6 RSB and Lévy Flight

Transition from paramagnetic to glassy states in random lasers has been also compared to the transition from diffusive to super diffusive (Lévy flight) regimes of light propagation. Lévy flight is a term used to describe the motion of light when it travels in a random series of shorter and longer steps differently from conventional Brownian diffusion characterized by the movement of light through random walks with same lengths of individual steps. In Lévy flights extremely long jumps can occur and dominate the transport of light through the material, that can become an optical super diffuser. Lévy random walks have power law distribution and super diffusion is a result of such long-tailed distribution, this is typical of physical phenomena with large fluctuations. As a demonstration, the heavy-tail aspect of the probability density of widely fluctuating emitted beam size and intensity in suitable colloidal samples have been shown and a net distinction between diffusive and super diffusive regimes has been illustrated (Barthelemy et al., 2008). Strong

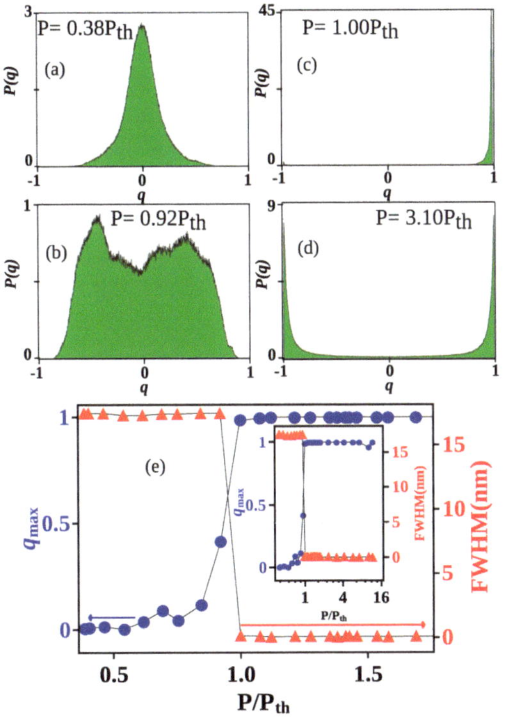

Figure 6.9 (a-d) P(q) obtained from experimental data at the indicated pump powers (normalized with respect to the threshold power P_{th}). (e) q_{max} (circles) versus normalized pump power, together with the FWHM (triangles) for comparison. The inset shows the results for pump powers up to $12P_{th}$. Reproduced from Ref. (Gomes et al., 2016a) with permission from American Physical Society.

emission fluctuations in RLs have been described in past by using the diffusive model of light and by identifying successfully different statistical regimes for disordered lasers (Lepri, 2013; Lepri et al., 2007). Within this theoretical approach the probability distribution of the emitted intensity is mainly dependent on the gain length l_g in units of the average mean free path of light $<l>$ inside the complex system, namely on the parameter $\alpha = l_g/<l>$. The authors nicely define a l_g versus L (length of the sample) phase diagram of the different diffusion regimes: For low pumping and high l_g the intensity fluctuations follow a Gaussian dispersion with $\alpha = 2$, by increasing gain and in a defined region below ($1 < \alpha < 2$) and above ($1/2 < \alpha \leq 1$) laser threshold the Lévy statistics of the emission fluctuations is evidenced. Well above threshold ($\alpha < 2$) weak fluctuations with Gaussian density distribution occur due to the depletion of the population inversion and the finite size of the active photonic system.

These intriguing findings have been recently considered in novel RLs, where the strong emission fluctuations have been analyzed both in the frameworks of Lévy statistics and spin glass theory. Interesting similarities and differences have been demonstrated.

A first comparison between the two statistical methods is reported by Gomes et al. in (Gomes et al., 2016b). In this work the physical origin of the complex correspondence between the Lévy flight statistics of emission intensity and the photonic RSB glassy transition in RLs is both theoretically and experimentally investigated . The RL consisted of crystalline powders of Nd^{3+}-doped YBO_3 (Nd:YBO) Nd^{3+} with 4.0 mol % concentration. The nanoparticle scatterers also act as the gain medium and the system is a solid-state material where scatterers positions do not change from shot to shot, in this way identical experimental conditions are given over the many subsequent excitation pulses. The powder is placed on a sample holder and gently pressed into a uniform disc region, then it is injected by an Optical Parametric Oscillator (OPO) pumped by the second-harmonic of a Q-switched Nd:YAG laser (7 ns, 10 Hz) at 806 nm wavelength leading to a RL emission peaking at 1056 nm. Over laser threshold, the spectra present one sharp sub-nanometric peak and, based on the same arguments of Ref. (Gomes et al., 2016a), the authors state that the emission is multi-modal with undistinguishable resonances because of the low instrumental resolution.

In this article, for the first time the authors explain the comparison between the RSB transition to the photonic glassy phase and the changes in the statistics of intensity fluctuations in RLs within the same theoretical framework based on the Langevin equations ruling the evolution of the mode amplitudes. While such theoretical approach has been introduced and widely used by the Spin Glass community (Angelani et al., 2006; Antenucci et al., 2015b; Leuzzi et al., 2009), only recently it has been considered for the statistical distribution of RL intensities (Raposo and Gomes, 2015). Starting from the same Langevin equation

$$\frac{\mathrm{d}a_k}{\mathrm{d}t} = \frac{\partial H}{\partial a_k^*} + F_k, \tag{6.11}$$

with $a_k(t)$ the slow-amplitude modes, compared to the rapidly evolving phase dynamics, H the complex-valued functional (hamiltonian of the system under specific assumptions) and F_k a Gaussian white noise, the Fokker–Planck for the probability density function (PDF) of emission intensity is obtained. The steady state solution of such equation gives the exponentially truncated Lévy-like distribution of intensities

$$P(I_k) = A\, I_k^{-\mu_k} \exp(-\gamma I_k), \quad \gamma > 0, \tag{6.12}$$

for $1 < \mu_k < 3$, and characterized by strong fluctuations of I_k. A and γ are constants. In contrast, for $\mu_k \leq 1$ or $\mu_k \geq 3$ the central limit theorem provides fast convergence to Gaussian statistics with $\alpha = \mu_k - 1 = 2$ and weakly fluctuating intensities. In practice, for a given disorder strength, an increasing pumping rate augments the value of $\mu_k = \alpha + 1$ and the statistics of emission intensities shifts progressively from an initial Gaussian ($\mu_k \leq 1$, $\alpha = 2$), to a Lévy-like ($1 < \mu_k < 3$, $0 < \alpha_k < 2$), and to a subsequent Gaussian ($\mu_k \geq 3$, $\alpha = 2$) regime, as predicted by the diffusive model of light propagations in random media.

Experimentally the authors report on the strong spectral fluctuations and analyze these by means of the two statistical theories. The results are shown in Figs. 6.10 and 6.11. It is clear that fluctuations are stronger close to the threshold (Figs. 6.10b,c), and from the evolution of $P(q)$ the RSB transition from the spontaneous emission-paramagnetic to the RL glassy behavior at threshold is evident both in Figs 6.10 and 6.11a. Correspondingly the sharp transition from Gaussian to Lévy statistics regime is demonstrated by the occurrence of $0 < \alpha < 2$. The novelty of these findings lies at higher excitation pulse energy well above the threshold, where fluctuations decline considerably with the consequent restoration of the Gaussian regime (Fig. 6.11b) and the partial suppression of spin glass behavior with $q_{\max} < 1$. In the article, the authors do not give any explanation of the existence of a strict causal link between the self-averaged Gaussian regime of intensity fluctuations above

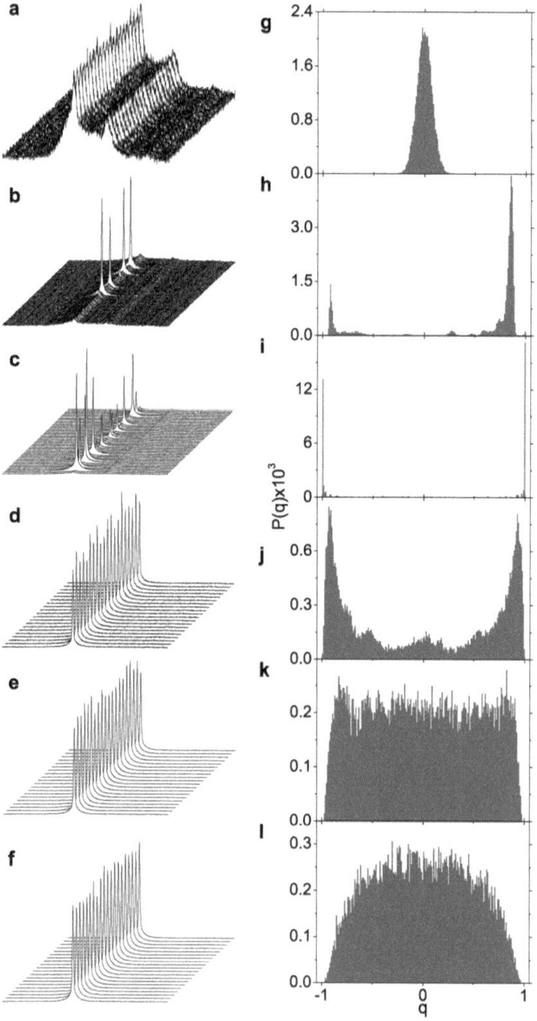

Figure 6.10 (a–f) Intensity spectra showing the fluctuations from shot to shot of the Nd:YBO system for excitation pulse energies (a) 1.20 mJ (below the RL threshold), (b) 1.36 mJ, (c) 1.4 mJ (both around the threshold), (d) 1.60 mJ, (e) 2.20 mJ and (f) 2.80 mJ (above the threshold). (g–l) Corresponding PDF distributions of the overlap parameter q. Reproduced from Ref. (Gomes et al., 2016b) with permission from Springer Nature Publishing AG.

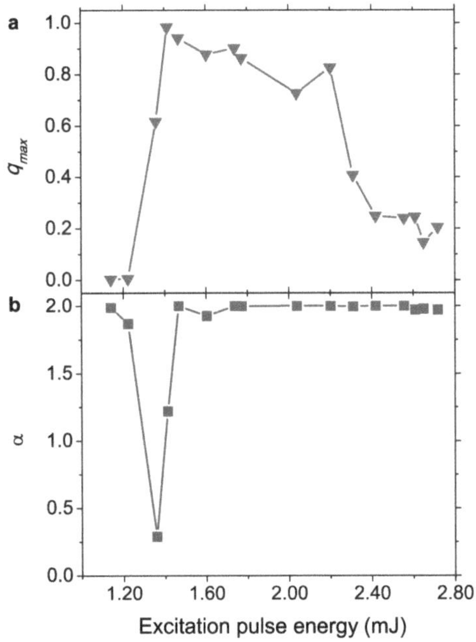

Figure 6.11 (a) The value $q_{\max}$ at which the Parisi overlap order parameter assumes its maximum. (b) The Lévy parameter α versus pumping energy. calculated from the data in Fig. 6.10. Reproduced from Ref. (Gomes et al., 2016b) with permission from Springer Nature Publishing AG.

the threshold and the observed trend to suppress the glassy phase should, leaving the subject to further studies.

A different result is shown in Ref. (Tommasi et al., 2016), where the authors conclude that the RSB and the Lévy regime onset may not have a clear causal relation between them. They claim that the former is strongly related to the threshold and less on the magnitude of the fluctuations, while the latter emerges for large fluctuations and only under strict conditions of the ratio between l_g (gain length) and $< l >$ (scattering mean free path). As a demonstration they report the clear experimental result reported in Fig. 6.12 where for large mean free path the Lévy regime is suppressed while RSB and glass transition is present. They write that a deeper study of this issue, both experimental and theoretical, is required.

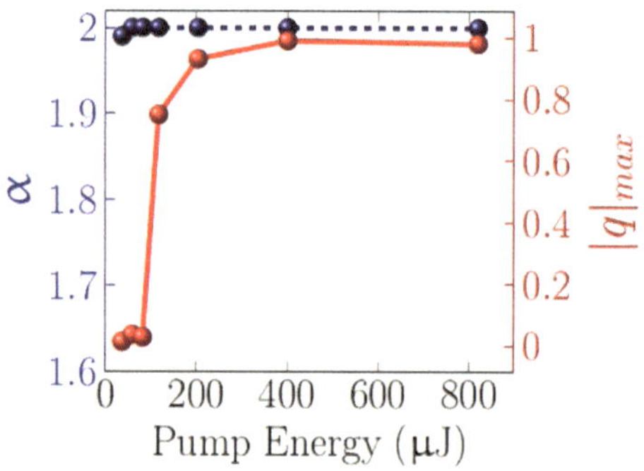

Figure 6.12 α and $|q|_{\max}$ indexes are shown. An abrupt change of $|q|_{\max}$ is found at threshold, independently on the presence of the Lévy regime ($\alpha < 1.8$). Reproduced from Ref. (Tommasi et al., 2016) with permission from Springer Nature Publishing AG.

The agreement between the onset of the Lévy statistical regime of intensity fluctuations and the emergence of the RSB glassy RL phase has been recently reported also in a one dimensional Er-RFL (Lima et al., 2017). It is clearly shown that both Gaussian-to- Lévy and spin glass transition in the statistics of intensity fluctuations are simultaneously present around the RL threshold. Similarly to the three dimensional systems (Gomes et al., 2016a,b) Gauss–Lévy–Gauss transitions in the statistics of the intensity fluctuations is shown: By increasing pumping, close to the threshold, a given mode can dominate over the others, retaining most of the gain. As a consequence, the possibility of extreme events characteristic of heavy-tailed power-law Lévy distributions arises notably and the probability of very long paths of diffusing photons with large gain is enhanced. On the other hand, as the input power further increases, the distribution of the gain among the active modes becomes more and more homogeneous, diminishing considerably the probability of extreme events of intensity and rather long diffusing paths. In this situation, the system enters a self-averaging gain regime and a Gaussian distribution of output intensities takes place above the threshold. In this work, the abrupt decrease of α is shown to be closely related to the sharp linewidth reduction at the onset of RL threshold. For higher pumping, after reaching a minimum around the threshold, α smoothly rises back to the Gaussian value achieved above the threshold. This implies the crossover towards a RL state with linewidth narrowing and Gaussian intensity fluctuations of intermediate strength between the prelasing and Lévy regimes.

6.7 RSB and Turbulence

In the last two years, the strong intensity fluctuations in random lasers have been also analyzed in the framework of turbulence theory (González et al., 2017, 2018). Interestingly, the emergence of turbulence behavior is demonstrated to coincide precisely with the onset of the photonic replica-symmetry-breaking glassy phase at the laser threshold in erbium random fiber laser (Er-RFL), in which a randomly distributed phase error grating is written. Using this procedure, a very high number of scatterers ($\gg$1000) is obtained, improving the fiber randomness. The length of the fiber used is 30 cm and the system is pumped by a CW laser. The RL threshold power from the analysis of the full width at half maximum (FWHM) is $P_{th} = (16.30 \pm 0.05)$ mW. The Er-RFL linewidth is limited by the instrumental resolution to 0.1 nm and the number of longitudinal modes, measured using a speckle contrast technique, is $\sim$204 confirming the multimode character of the Er-RFL system. In Ref. (González et al., 2017), the authors start from a dynamical hierarchical model (Macêdo et al., 2017) and adapt the basic concepts of Kolmogorov's statistical approach to turbulence. The main points are as follows: (i) Energy cascade: Energy is transferred from large to small scales. (ii) Intermittency: It is the tendency of the distribution of velocity differences between two points in the fluid to develop long non-Gaussian tails. In Kolmogorov's theory, it is assumed that at large Reynolds number big eddies are created spontaneously, which, because of large inertia effects, must decay into smaller eddies, in a cascade of events that go all the way down to the smallest scale, where all eddies disappear through viscous dissipation. In this way, intermittency is caused by fluctuations in the energy fluxes between adjacent scales; otherwise, in case of constant energy fluxes, the fluid velocity fluctuations would display regular Gaussian statistics.

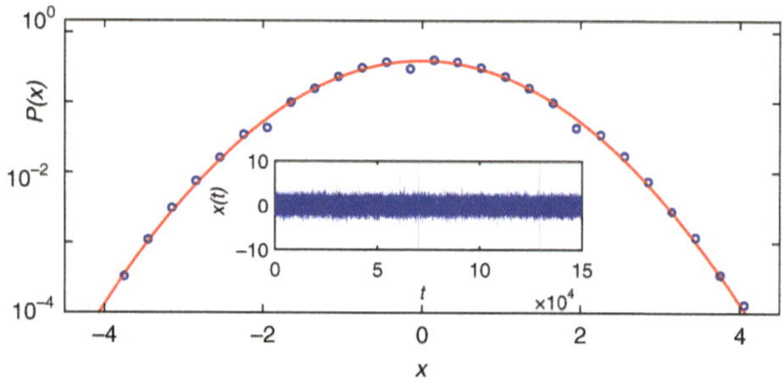

Figure 6.13 Semi-log plot of the distribution $P(x)$ of experimental intensity increments x in the non-turbulent spontaneous emission regime at the normalized excitation power $P/P_{\text{th}} = 0.72$ (blue circles) and best fit to a Gaussian expression (red line). The inset displays the corresponding time series of $x(t)$ as function of the dimensionless time index t. Reproduced from Ref. (González et al., 2017) with permission from Springer Nature Publishing AG.

Similar to the spin glass theory analysis, to investigate the fluctuations the emission spectral peak intensities are considered over 150,000 shots and in this case their time evolution is quantitatively analyzed.

The observable is the intensity increment $x(t) = I(t+1) - I(t)$ between successive optical spectra retained as stochastic parameter. In absence of nonlinearities, as in the prelasing regime, the intensity increments are statistically independent and the probability distribution $P(x)$ is a Gaussian. As the input power is increased beyond the threshold, dynamical nonlinearities lead to turbulent emission. This implies that the Gaussian form of the signal x distribution remains valid only at a local level, The nice results are depicted in Fig. 6.13 for the Gaussian non-lasing case and in Fig. 6.14 for emissions close (upper panel) and well above threshold (lower panel). Near laser threshold a single time scale in the dynamics of the background variable characterizes the turbulent behavior of the intensity fluctuations dynamics in Er-RFL, while the observed rise in the number of relevant time scales by increasing the excitation power in Fig. 6.14 is a clear indication of the multiscale turbulent cascade behavior of the dynamics of the intensity fluctuations.

In 2018, the same authors employed the same RFL system and same experimental techniques to demonstrate for the first time in

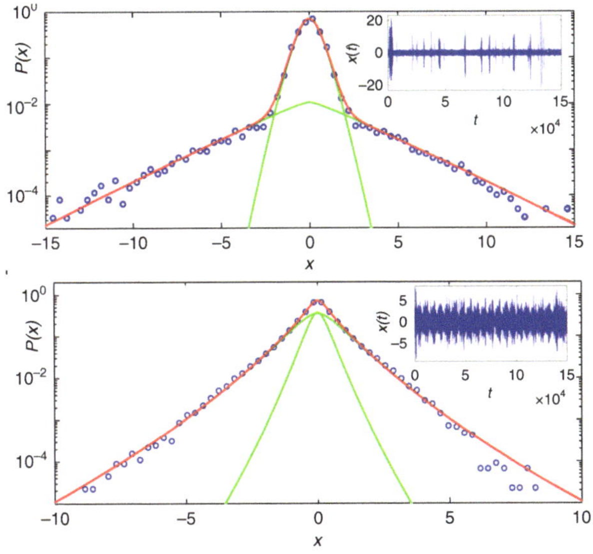

Figure 6.14 Semi-log plot of the distribution $P(x)$ of experimental intensity increments x (blue circles) near the threshold at the normalized excitation power $P/P_{\text{th}} = 0.99$ (upper panel) and above threshold for $P/P_{\text{th}} = 2.92$ (lower panel). Red lines are the predictions from the hierarchical model. The individual components of the mixture, given by two K-distributions for the upper panel and by two Meijer G-distributions for the lower one are evidenced by green lines. The insets show the corresponding time series of intensity increments. Reproduced from Ref. (González et al., 2017) with permission from Springer Nature Publishing AG.

any physical system the coexistence of turbulence-like and spin glass behavior from the same set of measurements (González et al., 2018).

For the theoretical analysis, they introduced the photonic Pearson correlation coefficient that can be considered as a generalized time τ dependent expression of the spin-glass Parisi overlap parameter:

$$Q_{\alpha\beta,\tau} = \frac{\sum_k \delta_{\alpha\tau}(k)\delta_{\beta\tau}(k)}{\sqrt{\sum_k \delta^2_{\alpha\tau}(k)}\sqrt{\sum_k \delta^2_{\beta\tau}(k)}}, \tag{6.13}$$

with k wavelength index in the emission spectra, α and β single-shot spectra indexes (replicas),

$$\delta_{\alpha\tau}(k) = y_{\alpha\tau}(k) - \bar{y}_\tau(k) \tag{6.14}$$

and

$$y_{\alpha\tau}(k) = I_{\alpha\tau}(k) - cI_{\alpha}(k). \tag{6.15}$$

Equation (6.14) determines the fluctuation around the average of a new variable $y_{\alpha}(k)$ defined in Eq. (6.15) as the temporal variation of the emitted intensity $I_{\alpha}(k)$. For $\tau = 0$ and $c = 0$, the time dependent fluctuation coincides with Eq. (6.8) as $\delta_{\alpha\tau}(k) = \Delta_{\alpha}(k)$ and the Pearson coefficient recovers the Parisi overlap as $Q_{\alpha\beta,\tau=0} = q_{\alpha\beta}$.

With these notations Eq. (6.13) quantifies the emission intensity fluctuations in time and it is sensitive to both liquid-glass transition and fluid dynamics phenomena such as turbulence.

In Fig. 6.15 from Ref. (González et al., 2017) experiments and theoretical analysis show that above threshold at short time $\tau = 1$ the maximum turbulence-like signature (intermittency) is manifested. This is demonstrated by the large deviation from the parabolic shape observed in this semi-log plot of Fig. 6.15b consistent with the non- Gaussian photonic turbulent-like behavior of $P(x_1)$. From same data in Fig. 6.15c it is demonstrated that Pearson coefficient for short times have their probability much increased around zero $Q_1 \sim 0$, leading to the unimodal behavior of $P(Q_1)$ in the turbulent-like state.

A different scenario is given for $\tau >> 1$ and above threshold as reported in Fig. 6.16. At the large time scale $\tau = 5000$, the parabolic shape of the distribution $P(x_{5000})$ in Fig. 6.16b assumes a Gaussian form showing that the short-time-scale effects and intermittency vanish and a crossover to the non-turbulent (Gaussian) behavior takes place.

The authors of Ref. (González et al., 2018) conclude that the unimodal behavior of $P(Q_1)$ above threshold at $P/P_{\text{th}} = 2.92$ should not be confused with the unimodality of $P(q)$, which, instead, signals the prelasing phase with replica symmetry for $P/P_{\text{th}} < 1$). Moreover, the bimodal behavior of $P(Q_{5000})$ resembles the spin-glass profile of the distribution $P(q)$ of the Parisi overlap parameter q for $P/P_{\text{th}} > 1$. They state that, since the Parisi overlap parameter considers all separation times τ between spectra, they expect that the statistical weight of the replica overlaps with $\tau >> 1$ dominates over the long time series. As a consequence, $P(Q_{\tau})$ actually appears qualitatively similar to the distribution $P(q)$ that characterizes the RSB spin glass phase.

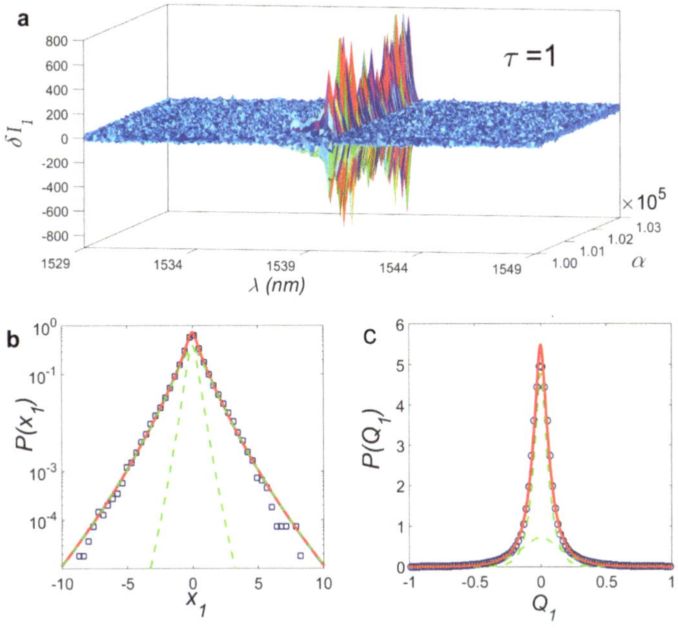

Figure 6.15 (a) Plot of 3000 experimental intensity increments $\delta I_{\alpha\tau}(k) = I_{\alpha+\tau}(k) - I_{\alpha}(k)$ between successive spectra for short times ($\tau = 1$) as a function of the spectrum label α and wavelength λ indexed by k. (b) Distribution $P(x_1)$ of normalized increments $x_{\alpha\tau}(k) = \delta I_{\alpha\tau}(k)/[\sum_{\alpha} \delta I_{\alpha\tau}^2(k)]^{1/2}$ at the emission peak, obtained from the whole set of 150,000 spectra (blue squares). The red solid line is the nice fit to the results of the hierarchical model, with $P(x_1)$ given by a statistical mixture of two Meijer-G functions. (c) Related distribution $P(Q_1)$ of the photonic Pearson coefficient Eq. (6.13). Reproduced from Ref. (González et al., 2018) with permission from Springer Nature Publishing AG.

These amazing findings can be explained by using the similarity with fluid dynamics and supercooled liquid transition phenomenology. Liquids above the glass transition temperature T_c (below laser threshold P_{th} in case of random lasers) have low viscosity and in this regime turbulence my take place, by decreasing temperature and through the liquid-glass transition the fluid viscosity grows drastically hampering turbulence. This is a static scenario, where the phases (liquid and glass) are retained at the stationary states. Things become rich and complex if dynamics is considered: Once below T_c,

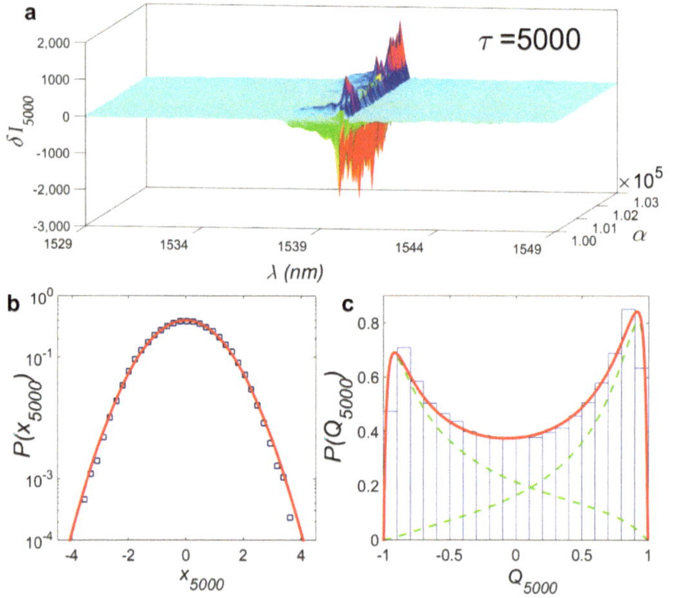

Figure 6.16 (a) Plot of 3000 experimental intensity increments, $\delta I_{\alpha\tau}(k)$, between spectra separated by $\tau = 5000$ time units, as a function of the spectrum label α and wavelength λ indexed by k. (b) Semi-log plot of the distribution $P(x_{5000})$ of normalized increments obtained from the whole set of 150,000 spectra (blue squares). The red solid line is the fit to the results of the hierarchical model. (c) Histogram with the distribution $P(Q_{5000})$ of the photonic Pearson coefficient calculated from 10,001 spectra displaying a spin-glass-like bimodal profile. The experimental histogram is well fitted by the theoretical model (red solid line). Reproduced from Ref. (González et al., 2018) with permission from Springer Nature Publishing AG.

the liquid-glass transition takes time to occur and at very short time the system is still in the liquid (paramagnetic) phase with Gaussian $P(Q)$, similar to the photonic case of Fig. 6.15c, and also to $P(q)$, if calculated at a precise very short time. In such dynamical regime turbulence can be present and the example of disordered RFL is well suited as shown in Fig. 6.15b.

After a certain time, that in glass transition theory is commonly name τ_{d} (Castellani and Cavagna, 2005; Debenedetti and Stillinger, 2001), the transition to an arrested out of equilibrium state is completed, the system is effectively in a glassy phase with bimodal

$P(Q)$ and $P(q)$, as nicely shown for the photonic case in Fig. 6.16c. In this regime turbulence is broken down by the *infinite* viscosity like in Fig. 6.16b.

6.8 RSB in Ordered Cavity

Before concluding the chapter on experiments in replica symmetry breaking in random laser, it is relevant to mention two recent works where RSB has been observed in standard ordered cavities.

Basak et al. in 2016 report on strong intensity fluctuations in both liquid and solid state dye lasers with Fabry–Perot cavities obtained by the cuvette walls in the former and by the interface of DNA-CTMA complex thin slab with air in the latter one (Basak et al., 2016). In both cases DCM dye is used as gain medium, this is dispersed in tetrahydrofuran for the liquid dye laser and in the DNA-CTMA matrix for the solidified one. The samples are then pumped with ns or ps pulsed Q-switched lasers. By increasing the pumping energy, sharp spectral line narrowing and peak intensity growth in the emission prove the transition from spontaneous to stimulated emission and at this threshold strong intensity fluctuations are observed. These are analyzed by means of the spin glass theory and RSB at laser threshold is clearly demonstrated.

For the authors the reason of this intriguing observation of replica symmetry breaking in absence of matter disorder resides in the immense number of modes and the reduced resonator quality causing the coupling of modes. The many (about 10^{10} mm^{-3}) modes exchange energy and interact but some of these interactions are frustrated due to cavity losses, in this way the fluctuations are explained in terms of mode coupling and frustration.

We add that this correct conclusion is very insightful but not unexpected. RSB has been experimentally proved in random lasers (Ghofraniha et al., 2015) and then also generalized to random waves (Pierangeli et al., 2017) and we think that the meaning of disorder in the analogy between random photonics and spin or structural glasses should be clarified. In spin glasses, disorder is given by dis-alignment of spin phases that provokes frustration in case of mutual interaction (coupling) at low temperature. In supercooled liquids and in colloidal glasses the spatial position of the molecules for the latter and of the macro-molecules in the former ones, that are not in crystalline ordered disposition, gives the disorder. Below a certain critical temperature, structural disorder together with the mutual interaction between molecules/colloids frustrates the system to relax to an equilibrium physical solid phase and leave it in a metastable non ergotic state. In random lasers, disorder is not the structural disorder of the material behaving as cavity, but is the mis-alignment between the phases of the electromagnetic emitted modes. This can be present also in ordered resonators, the crucial ingredients are the large number of the modes and their mutual coupling mediated by gain competition. In a perfect ordered cavity all modes are in phase and oscillate coherently, thus RSB should not be observed as shown in Fig. 6.7a,b and also in some of the cited papers. In case of slight misalignment of the laser resonator the active modes can oscillate in disordered way leading, similarly to RLs, to large intensity fluctuations with evidences of glassy behavior and RSB as nicely demonstrated in (Basak et al., 2016).

Different results are given in (Moura et al., 2017) where replica symmetry breaking is shown in the spontaneous mode-locking regime of a multimode Q-switched Nd:YAG laser. The authors assert that the observed phenomenon is quite distinct from what investigated in random lasers characterized by incoherently oscillating modes. Instead, in their case in each pulse a specific subset of longitudinal modes are activated in a nondeterministic way, whose coherent oscillation dominates and frustrates the others. The emergence of RSB coincides with the onset of ultrashort ps pulse generation typical of the spontaneous mode-locking regime, both occurring at the laser threshold. They also show that, when high losses are introduced, RSB is inhibited and only the amplified stimulated emission with replica symmetry is observed.

In this work, it is not clear what is the underlying mechanism of the reported glassy behavior and what is the frustration causing the illustrated strong emitted intensity fluctuations. We think that also in this investigation slight misalignment of the laser resonator could be responsible of the lack of perfect coherence of the modes. The large number of the active modes and energy competition provoke intensity fluctuations with glass-like transition at threshold. The same mechanisms are responsible of spontaneous mode-locking as shown by Moura et al.

6.9 RSB maps

Very recently, RSB theory has been applied to obtain the real time maps of the laser activity in a heterogeneous RL (Massaro et al., 2021). The random lasers are made of polymeric ribbon-like and highly porous fibers with evident RL action from separated micrometric domains that alternatively switch on and off by tuning the pumping light intensity. This novel effect is visualized by building for the first time replica symmetry breaking maps of the emitting fibers with 2 μm spatial resolution. The authors investigate this intriguing behavior by recording 100 single shot images at different input energies, and estimate for each pixel the mean intensity and the variance. The correspondent obtained maps at various pumping are shown in Fig. 6.17. At low energy (left panel in Fig. 6.17a), the spontaneous emission is homogeneously distributed along the fibers and does not present spatial heterogeneity in the fluctuations, as shown from the variance map. By increasing pumping, the average emission starts to be different in diverse points, and the regions with higher emission intensity also present high intensity fluctuations, this can be seen in Fig. 6.17a at 0.33 μJ.

For higher injected energy, a different behavior is observed: some domains of the active network have high mean emission intensity and high variance, while others, despite the strong brightness, are less fluctuating and more stable in time. In Fig. 6.17b, variance and mean intensity from two selected domains, D1 and D2, are plotted. The two regions present a growing mean emission by increasing the input energy, while the shot-to-shot fluctuations are divergent after 7.5 μJ. Both D1 and D2 are intense, but D2 is much more blinking at certain input energies. In Fig. 6.17b, plots of mean intensity and variance averaged over all pixels of the images are reported. Both present a growing trend for increasing pump showing that the emission from single parts of the network does not follow that from the whole system. To clarify these effects and for a visualization

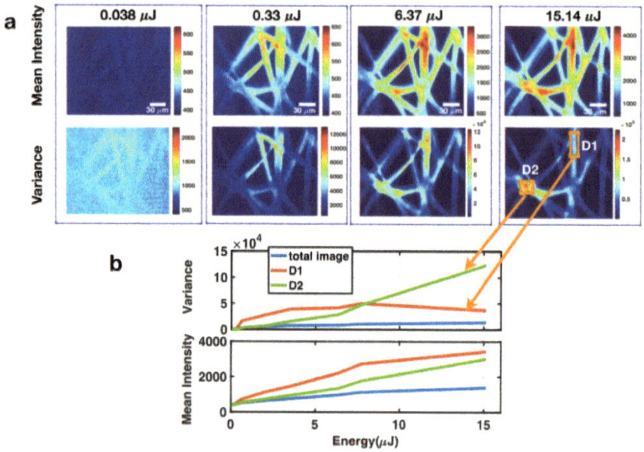

Figure 6.17 Maps of emitted mean intensity and variance. (a) Mean intensity and variance calculated from 100 single shot frames and mapped for various input laser energies. (b) Energy trends of the mean intensity and variance averaged over entire maps of panel a (blue line) and over two distinct domains, D1 and D2, indicated in panel a with orange squares. Reproduced from Ref. (Massaro et al., 2021) with permission.

of their activity, for the first time RSB maps from the fluorescence images are built. The overlap parameter is calculated directly from the images. One hundred single-shot frames are recorded at fixed energy, and each of them is a replica. Each frame of 1024 × 1376 pixels is divided into squares of 16 × 16 pixels, which are named macropixels in this work, and the overlap $q^j_{\alpha,\beta}$ for each macropixel is calculated as follows:

$$q^j_{\alpha,\beta} = \frac{\sum_{x_k}\sum_{y_k}\Delta_{\alpha,j}(x_k,y_k)\Delta_{\beta,j}(x_k,y_k)}{\sqrt{\sum_{x_k}\sum_{y_k}\Delta_{\alpha,j}(x_k,y_k)^2}\sqrt{\sum_{x_k}\sum_{y_k}\Delta_{\beta,j}(x_k,y_k)^2}}, \tag{6.16}$$

where α and β are replica indexes, j is the macropixel index, and x_k and y_k are pixel indexes inside one macropixel running from 1 to 16. In Eq. 6.16,

$$\Delta_{\alpha,j}(x_k,y_k) = I^j_\alpha(x_k,y_k) - I^{-j}_\alpha(x_k,y_k) \tag{6.17}$$

with $I^j_\alpha(x_k,y_k)$ being the intensity at the pixel with coordinate (x_k, y_k) of j-th macropixel for replica α and

$$I^{-j}_\alpha(x_k,y_k) = \frac{1}{N_{shot}}\sum_{\alpha=1}^{N_{shot}} I^j_\alpha(x_k,y_k) \tag{6.18}$$

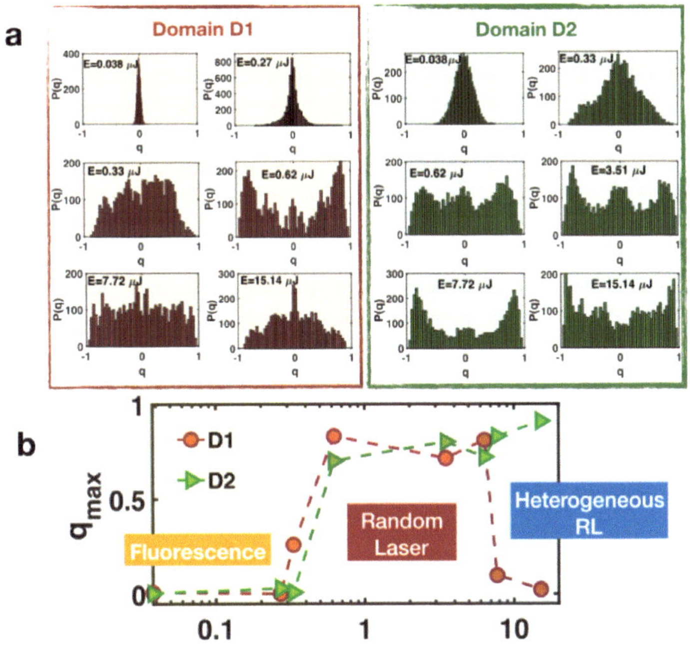

Figure 6.18 Overlap distributions evidencing heterogeneous RL. (a) Distributions of the overlap parameter q at various input energies and for two distinct domains D1 and D2, indicated in Fig. 6.17. (b) $q_{\max}$ vs. pump energy for D1 and D2 regions, showing three diverse regimes of the emission. Reproduced from Ref. (Massaro et al., 2021) with permission.

is the intensity at the pixel with coordinate (x_k, y_k) of j-th macropixel averaged over all 100 replica (shots). Macropixels of 16 × 16 pixels are chosen because this results in a good range for the sums in Eq. 6.16 and the estimation of $q^j_{\alpha,\beta}$ and it allows to have a good resolution in the RSB maps reported in the following.

By this procedure from the images, the authors have access to the sets of all $N_{\text{shot}}(N_{\text{shot}}1)/2$ values q^j of $q^j_{\alpha,\beta}$ and they determine the distributions $P(q^j)$ for all 64 × 86 macropixels j and for each pumping energy. Representative $P(q)$ distributions at different energies and for two macropixels inside the domains D1 and D2 of Fig. 6.17 are illustrated in Fig. 6.18.

At low energy, the distributions are centered around $q = 0$, meaning that both regions are in the paramagnetic regime of the spontaneous emission. By increasing pumping, the distributions broaden, and after a certain threshold, bimodal shapes around the values $q = \pm 1$ are obtained, signaling the breaking of replica symmetry and the RL out of equilibrium regime. At high pumping, a different and novel scenario emerges. In the macropixel inside region D1, the distribution $P(q)$ loses the bimodality, while the one in D2 does not present significant change.

Following the previous work, the authors choose $q_{\max}$, which is the absolute value of q for which $P(q)$ is maximum, as a suitable parameter to define the different regimes and transitions. In Fig. 6.18-b plots of $q_{\max}$ versus input energy for the macropixels inside domains D1 and D2 and estimated from the correspondent distributions are reported. The transition from spontaneous to stimulated emission at the RL threshold energy $E_{\text{th}} \cong 0.4$ μJ is defined by the value $q_{\max} = 0.5$.

At higher energy ($E \geq 7$ μJ), an unexpected divergence in $q_{\max}$ profile is clearly shown. The point with $q_{\max} \cong 1$ continues to lase, while the other with $q_{\max} \cong 0$ switches off. By calculating $q_{\max}$ for all macropixels of the fluorescence images replica symmetry breaking maps are obtained at various pumping and shown in Fig. 6.19. At low energy, $q_{\max} \leq 0.06$ and varies around the zero value in all points, and the whole system is in the fluorescence paramagnetic regime. At 0.33 μJ, a nucleation center of RL emission, defined by $q_{\max} \cong 0.6$, takes place. Above this energy other different parts of the heterogeneous RL are switched on and their activity changes randomly at different pumps. On the maps some of the ON ($q_{\max} > 0.5$) and OFF ($q_{\max} < 0.5$) regions are indicated, and their dynamics is clearly evidenced. The RSB maps in Fig. 6.19 show unambiguously which parts of the network have broken replica symmetry and are in the glassy-RL state and which parts have only spontaneous emission at distinct pumpings. In this work, the mapping of $q_{\max}$, that is, a parameter that can assume values between 0 and 1 and defines the transition to the laser regime, allows for the first time the visualization of heterogeneous RL with switching and variable activity with 2 μm spatial resolution given by the length of a macropixel. This procedure is proposed as a robust tool to identify the presence and the spatial extension of RL activity from fluorescence images.

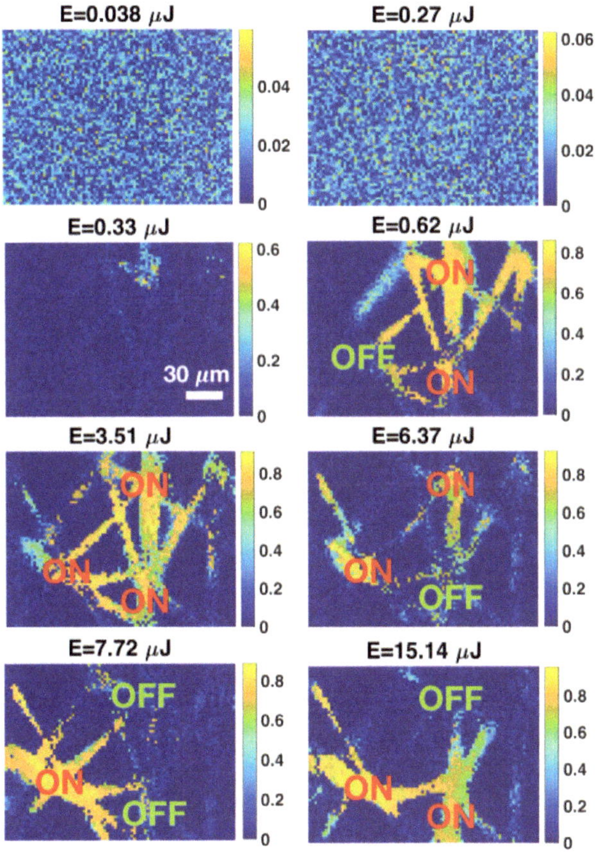

Figure 6.19 Replica symmetry breaking maps. Maps of the q_{max} values calculated for each macropixel at different input energies. ON and OFF indicate the activation status of the RL emission. Reproduced from Ref. (Massaro et al., 2021) with permission.

References

Angelani, L., Conti, C., Ruocco, G., and Zamponi, F. (2006). Glassy behavior of light, *Phys. Rev. Lett.* **96**, p. 065702.

Anni, M., Lattante, S., Stomeo, T., Cingolani, R., Gigli, G., Barbarella, G. and Favaretto, L. (2004). Modes interaction and light transport in bidimensional organic random lasers in the weak scattering limit, *Phys. Rev. B* **70**, p. 195216.

Antenucci, F., Conti, C., Crisanti, A., and Leuzzi, L. (2015a). General phase diagram of multimodal ordered and disordered lasers in closed and open cavities, *Phys. Rev. Lett.* **114**, p. 043901.

Antenucci, F., Conti, C., Crisanti, A., and Leuzzi, L. (2015b). General phase diagram of multimodal ordered and disordered lasers in closed and open cavities, *Phys. Rev. Lett.* **114**, p. 043901.

Antenucci, F., Crisanti, A., and Leuzzi, L. (2015c). The glassy random laser: Replica symmetry breaking in the intensity fluctuations of emission spectra, *Sci. Rep.* **5**, p. 16792.

Barthelemy, P., Bertolotti, J., and Wiersma, D. S. (2008). A Lévy flight for light, *Nature* **453**, p. 495.

Basak, S., Blanco, A., and López, C. (2016). Large fluctuations at the lasing threshold of solid- and liquid-state dye lasers, *Sci. Rep.* **6**, p. 32134.

Cao, H., Jiang, X., Ling, Y., Xu, J. Y., and Soukoulis, C. M. (2003). Mode repulsion and mode coupling in random lasers, *Phys. Rev. B* **67**, p. 161101.

Castellani, T., and Cavagna, A. (2005). Spin glass theory for pedestrians, *arXiv:cond-mat/0505032* .

Conti, C., and Leuzzi, L. (2011). Complexity of waves in nonlinear disordered media, *Phys. Rev. B* **83**, p. 134204.

de Matos, C. J. S., de S. Menezes, L., Brito-Silva, A. M., Martinez Gámez, M. A., Gomes, A. S. L., and de Araújo, C. B. (2007). Random fiber laser, *Phys. Rev. Lett.* **99**, p. 153903.

Debenedetti, P. G., and Stillinger, F. H. (2001). Supercooled liquids and the glass transition, *Nature* **410**, p. 259.

Fallert, J., Dietz, R. J. B., Sartor, J., Schneider, D., Klingshirn, C., and Kalt, H. (2009). Co-existence of strongly and weakly localized random laser modes, *Nat. Photon.* **3**, p. 279.

Frolov, S. V., Vardeny, Z. V., Yoshino, K., Zakhidov, A., and Baughman, R. H. (1999). Stimulated emission in high-gain organic media, *Phys. Rev. B* **59**, pp. R5284–R5287.

Ghofraniha, N., Viola, I., Dimaria, F., Barbarella, G., Gigli, G., Leuzzi, L. and Conti, C. (2015). Experimental evidence of replica symmetry breaking in random lasers, *Nat. Comm.* **6**, p. 6058.

Gomes, A. S. L., Lima, B. C., Pincheira, P. I. R., Moura, A. L., Gagné, M., Raposo, E. P., de Araújo, C. B., and Kashyap, R. (2016a). Glassy behavior in a one-dimensional continuous-wave erbium-doped random fiber laser, *Phys. Rev. A* **94**, p. 011801.

Gomes, A. S. L., Raposo, E. P., Moura, A. L., Fewo, S. I., Pincheira, P. I. R., Jerez, V., Maia, L. J. Q., and de Araújo, C. B. (2016b). Observation of Lévy distribution and replica symmetry breaking in random lasers from a single set of measurements, *Sci. Rep.* **6**, p. 27987.

González, I. R. R., Lima, B. C., Pincheira, P. I., Brum, A. A., Macêdo, A. M., Vasconcelos, G. L., de S. Menezes, L., Raposo, E. P., Gomes, A. S., and Kashyap, R. (2017). Turbulence hierarchy in a random fibre laser, *Nat. Comm* **8**, p. 15731.

González, I. R. R., Raposo, E. P., Macêdo, A. M., de S. Menezes, L., and Gomes, A. S. (2018). Coexistence of turbulence-like and glassy behaviours in a photonic system, *Sci. Rep.* **8**, p. 17046.

Leonetti, M., Conti, C., and López, C. (2013). Dynamics of phase-locking random lasers, *Phys. Rev. A* **88**, p. 043834.

Lepri, S. (2013). Fluctuations in a diffusive medium with gain, *Phys. Rev. Lett.* **110**, p. 230603.

Lepri, S., Cavalieri, S., Oppo, G., and Wiersma, D. (2007). Statistical regimes of random laser fluctuations, *Phys. Rev. A* **75**, p. 063820.

Leuzzi, L., Conti, C., Folli, V., Angelani, L., and Ruocco, G. (2009). Phase diagram and complexity of mode-locked lasers: From order to disorder, *Phys. Rev. Lett.* **102**, p. 083901.

Lima, B. C., Gomes, A. S. L., Pincheira, P. I. R., Moura, A. L., Gagné, M., Raposo, E. P., de Araújo, C. B., and Kashyap, R. (2017). Observation of Lévy statistics in one-dimensional erbium-based random fiber laser, *J. Opt. Soc. Am. B* **34**, 2, pp. 293–299.

Macêdo, A. M. S., González, I. R. R., Salazar, D. S. P., and Vasconcelos, G. L. (2017). Universality classes of fluctuation dynamics in hierarchical complex systems, *Phys. Rev. E* **95**, p. 032315.

Massaro, L. M., Gentilini, S., Portone, A., Camposeo, A., Pisignano, D., Conti, C. and Ghofraniha, N. (2021). Heterogeneous Random Laser with Switching Activity Visualized by Replica Symmetry Breaking Maps, *ACS Photon.* **8**, pp. 376–383.

Mézard, M., Parisi, G., and Virasoro, M. (1987). *Spin Glass Theory and Beyond* (World Scientific, Singapore).

Moura, A. L., Pincheira, P. I. R., Reyna, A. S., Raposo, E. P., Gomes, A. S. L. and de Araújo, C. B. (2017). Replica symmetry breaking in the photonic ferromagneticlike spontaneous mode-locking phase of a multimode Nd:YAG laser, *Phys. Rev. Lett.* **119**, p. 163902.

Mujumdar, S., Türck, V., Torre, R., and Wiersma, D. S. (2007). Chaotic behavior of a random laser with static disorder, *Phys. Rev. A* **76**, p. 033807.

Parisi, G. (1980). The order parameter for spin glasses: A function on the interval 0-1. *J. Phys. A* **13**, p. 1101.

Parisi, G. (1983). Order parameter for spin-glasses, *Phys. Rev. Lett.* **50**, pp. 1946–1948.

Pierangeli, D., Tavani, A., Mei, F. D., Agranat, A. J., Conti, C., and DelRe, E. (2017). Observation of replica symmetry breaking in disordered nonlinear wave propagation, *Nat. Comm.* **8**, p. 1501.

Pincheira, P. I. R., Silva, A. F., Fewo, S. I., no, S. J. M. C., Moura, A. L., Raposo, E. P., Gomes, A. S. L., and de Araújo, C. B. (2016). Observation of photonic paramagnetic to spin-glass transition in a specially designed TiO_2 particle-based dye-colloidal random laser, *Opt. Lett.* **41**, 15, pp. 3459–3462.

Raposo, E. P., and Gomes, A. S. L. (2015). Analytical solution for the Lèvy-like steady-state distribution of intensities in random lasers, *Phys. Rev. A* **91**, p. 043827.

Skipetrov, S. E., and Maynard, R. (2000). Instabilities of waves in nonlinear disordered media, *Phys. Rev. Lett.* **85**, pp. 736–739.

Tommasi, F., Ignesti, E., Lepri, S., and Cavalieri, S. (2016). Robustness of replica symmetry breaking phenomenology in random laser, *Sci. Rep.* **6**, p. 37113.

Uppu, R., and Mujumdar, S. (2010). Statistical fluctuations of coherent and incoherent intensity in random lasers with nonresonant feedback, *Opt. Lett.* **35**, 17, pp. 2831–2833.

van der Molen, K. L., Mosk, A. P., and Lagendijk, A. (2006). Intrinsic intensity fluctuations in random lasers, *Phys. Rev. A* **74**, p. 053808.

Chapter 7

Turbulence-Like Phenomena in Random Lasers, Coexistence with a Photonic Spin-Glass Phase and Modes Correlation through Pearson Statistics

Antônio M. S. Macêdo,[a] Giovani L. Vasconcelos,[b] Ivan R. R. Gonzalez,[a] Anderson S. L. Gomes,[c] and Ernesto P. Raposo[a]

[a] *Laboratório de Física Teórica e Computacional, Departamento de Física, Universidade Federal de Pernambuco, 50670-901 Recife, Pernambuco, Brazil*
[b] *Departamento de Física, Universidade Federal do Paraná, 81531-990 Curitiba, Brazil*
[c] *Departamento de Física, Universidade Federal de Pernambuco, 50670-901 Recife, Pernambuco, Brazil*
antonio.smacedo@ufpe.br

7.1 Introduction

Turbulence has been observed in many natural systems, such as the atmosphere, the oceans, and a variety of biological environments [1–4], as well as in artificial systems, which include fiber lasers [5, 6], Bose–Einstein condensates [7], nonlinear optics [8], and integrable

Lévy Statistics and Spin Glass Behavior in Random Lasers
Edited by Anderson S. L. Gomes, André L. Moura, Cid B. de Araújo, and Ernesto P. Raposo

ISBN 978-981-4968-55-3 (Hardcover), 978-1-003-33618-1 (eBook)
www.jennystanford.com

systems and solitons [9, 10]. In a broader context, turbulence theory has also been used to explain financial market features [11, 12]. In particular, atmospheric turbulence impacts optical communications, in which the encoding of information in orbital angular momentum has been employed as a form of mitigation, in a context where turbulence plays a relevant role [13].

Two crucial ingredients of fluid turbulence are (i) the energy cascade whereby energy is transferred from larger to smaller scales, until it is dissipated by viscous forces, and (ii) the phenomenon of intermittency, which results from fluctuations of the rate of energy transfer across this hierarchy of scales. At highly turbulent flows, i.e., for large Reynolds number, big eddies are created spontaneously, which, because of large inertia effects, decay into smaller eddies, in a cascade of events that go all the way down to the smallest scale, where all eddies disappear through viscous dissipation [14]. In this view, intermittency is then caused by fluctuations in the energy transfer rates or energy fluxes between adjacent scales, because otherwise (that is, if the energy fluxes were constant) the fluid velocity fluctuations would display regular Gaussian statistics. It turns out that this hierarchical description of turbulence can also be applied to other non-Gaussian fluctuation phenomena, as random lasers (RLs).

In contrast to both conventional and fiber lasers, where the feedback providing gain amplification is mediated by a closed cavity formed by mirrors or fiber Bragg reflectors [15, 16], the optical feedback in RLs and random fiber lasers (RFL) arises from the multiple scattering of photons in a disordered medium, therefore forming an open complex disordered nonlinear system, in which light propagation occurs in the presence of gain, leading to laser emission [17, 18]. It is precisely this "flow of photons" in a disordered nonlinear active medium that gives rise to turbulent-like behavior in the intensity emission of RL systems.

In this chapter, we start by describing a theoretical framework, called H-theory, that has been used to describe the turbulent-like emission in an erbium-doped RFL (Er-RFL) with continuous-wave (CW) pump. This formalism is based on a multiscale approach to hierarchical complex dynamical systems and has been applied with success, for instance, to both fluid turbulence [19] and RFL [20].

Turbulent (or perhaps more appropriately termed turbulent-like or turbulent-type) behavior in an RFL has been observed to occur both near and above the laser threshold transition [20]. In the turbulent emission regime, the probability distribution of the intensity fluctuations can be described analytically by a mixture of statistics that arise from a hierarchical stochastic model of intermittency, as discussed below.

We also present in this chapter a joint analysis of the photonic turbulence-like, spin-glass, and mode correlations properties through the introduction of a couple of Pearson correlation coefficients. We finally argue in the concluding section on the possibility to understand these complex behaviors under a unified theoretical approach that can benefit from tools and concepts employed in their separate treatment.

7.2 Hierarchical Stochastic Model

Consider a stationary complex system with multiple time scales: $\tau_0 \gg \tau_1 \gg \cdots \gg \tau_N$. Suppose that the system is observed via the time series $\{x_\tau(t)\}$ of the increments of some macroscopic variable $\theta(t)$, where $x_\tau(t) = \theta(t+\tau) - \theta(t)$. For example, in turbulence $x(t)$ is the fluid velocity increments $x_\tau(t) = v(t+\tau) - v(t)$, whereas in RLs we shall consider the intensity increments between successive optical spectra, i.e., $x_1(t) \equiv I(t+1) - I(t)$; see below. The largest time scale τ_0 in the system is usually related to dynamical structures spanning the system size, while the parameters τ_i, $i = 1, \ldots, N$, correspond to the time scales of coherent structures of smaller sizes present in the system. For convenience of notation, in the remainder of this section the increments measured at a time scale τ_i will be labeled simply by the order i within the hierarchy of time scales, i.e., $x_i(t) \equiv x_{\tau_i}(t)$.

We assume that at very large time scales, i.e., at a time scale $\tau \gtrsim \tau_0$, the increments x_0 are uncorrelated, so that the large-scale probability distribution $P(x_0|\epsilon_0)$ can be seen as the *global equilibrium distribution* of the system, where the parameter ϵ_0 characterizes the system's large-scale stationary state. For example, in turbulence ϵ_0 is related to the rate of energy injected at the large scales by the forcing mechanism driving the turbulent flow. To be specific, here we assume that $P(x_0|\epsilon_0)$ is a Gaussian distribution with zero mean and variance ϵ_0:

$$P(x_0|\epsilon_0) = \frac{1}{\sqrt{2\pi\epsilon_0}} \exp\left[-\frac{x_0^2}{2\epsilon_0}\right]. \tag{7.1}$$

In many complex systems, such as turbulent flows and RLs, the observed distributions at small scales deviate significantly from the expected Gaussian, with the small-scale distributions often exhibiting heavy tails—the smaller the scale, the heavier the tail. This dependence of the shape of the distributions on the scale in which the observations are made is sometimes referred to as *intermittency*. Here we describe a theoretical framework in which this intermittency effect naturally emerges as the result of the complex multiscale dynamics of the system.

Suppose that the measurements of interest are made at some short time scale $\tau \ll \tau_0$. We assume that τ corresponds to the N-th level in the system's hierarchy of time scales, i.e., $\tau \sim \tau_N$, where N can be estimated as $N = \ln_2(\tau_0/\tau)$. According to the convention above, the increments measured at this scale will be denoted by $x_N(t)$. In view of the separation of time scales, the observed quantity $x_N(t)$ is supposed to fluctuate much more rapidly than the larger dynamical structures (hence with larger characteristic times) where the dynamics of x_N is "embedded." In other words, the larger structures can be viewed as providing a slowly varying "local environment" for the quantity x_N. This means that over sufficiently short time intervals, during which the environment has not had time to vary appreciably, the observed quantities x_N follow essentially the same large-scale distribution $P(x_N|\epsilon_N)$, but with a new parameter ϵ_N that characterizes the "local equilibrium." However, owing to the system's multiscale complex dynamics, the local-equilibrium parameter ϵ_N also varies, albeit slowly, so that ϵ_N can be regarded as a (slow) random variable.

Thus, if we sample the system at short intervals but over a long time span during which the local variable ϵ_N also fluctuates, then the distribution of x_N will be described by the compound distribution

$$P(x_N) = \int_0^\infty P(x_N|\epsilon_N) f(\epsilon_N) d\epsilon_N, \tag{7.2}$$

where $f(\epsilon_N)$ is the stationary probability density function of the background variable ϵ_N. In the compound approach above, also called superstatistics [21], the complex statistical properties of the system is fully encoded in the weighting density $f(\epsilon_N)$, which incorporates the effect of the fluctuating background (intermittency). Recently, a new theoretical framework, called H-theory [22, 23], has been introduced to derive the distribution $f(\epsilon_N)$ from a hierarchical stochastic dynamical model. One distinctive aspect of H-theory is that one can obtain the background distribution $f(\epsilon_N)$ from general physical arguments, instead of simply prescribing it, as is normally done in the superstatistics formalism [21].

The exact expression for the background density is obtained in H-theory from the stationary solution of the following set of stochastic differential equations:

$$d\epsilon_i(t) = -\gamma_i\,(\epsilon_i - \epsilon_{i-1})\,dt + \kappa_i\sqrt{\epsilon_i\epsilon_{i-1}}dW_i(t), \qquad (7.3)$$

for $i = 1, \ldots, N$. The variable ϵ_i represents the fluctuating variance parameter at the respective scale i in the hierarchy, γ_i and κ_i denote positive constants, and $W_i(t)$ are independent Wiener processes. We recall that the Wiener process $W(t)$ is defined such that the increments $dW(t) = W(t + dt) - W(t)$ are Gaussian random variables with zero mean, $\langle dW(t)\rangle = 0$, and variance given by $\left\langle [dW(t)]^2\right\rangle = dt$.

The first term in Eq. (7.3) describes the deterministic coupling between adjacent scales, which leads to the relaxation (in average) towards ϵ_0:

$$\lim_{t\to\infty}\langle\epsilon_i(t)\rangle = \epsilon_0, \quad \forall i. \qquad (7.4)$$

In contradistinction, the stochastic (noise) term in (7.3) accounts for the nonlinear couplings with the background variables at all scales. It represents the source of intermittency, since if we omit this term (i.e., if we set $\kappa_i = 0$) we simply get a nonfluctuating equilibrium regime: $\lim_{t\to\infty}\epsilon_i(t) = \epsilon_0$. We remark furthermore that the form of Eq. (7.3) is dictated by three physical requirements: (i) the global equilibrium condition given in Eq. (7.4); (ii) the invariance under the scale transformation $\epsilon_i \to \lambda\epsilon_i$; and iii) the positivity requirement, i.e., $\epsilon_i(t) \geq 0$ with probability one for all $t > 0$, which requires that the noise amplitude vanish for $\epsilon_i = 0$; plus some additional constraint on the moments of the stationary distribution [23].

7.2.1 Stationary Solution

Under the assumption of large time scales separation, $\gamma_N \gg \gamma_{N-1} \gg \ldots \gg \gamma_1$, where $\gamma_i = \tau_i^{-1}$, we can "decouple" Eq. (7.3) as follows: For each level i, we keep the slower variable ϵ_{i-1} fixed and solve for the stationary conditional probability density $f(\epsilon_i|\epsilon_{i-1})$. More specifically, for each SDE given in Eq. (7.3) one writes down the associated Fokker–Planck equation for the probability distribution $f(\epsilon_i|\epsilon_{i-1}, t)$:

$$\frac{\partial f(\epsilon_i|\epsilon_{i-1}, t)}{\partial t} = \gamma_i \frac{\partial[(\epsilon_i - \epsilon_{i-1})\, f(\epsilon_i|\epsilon_{i-1}, t)]}{\partial \epsilon_i} + \frac{\kappa^2}{2} \frac{\partial^2[\epsilon_i \epsilon_{i-1} f(\epsilon_i|\epsilon_{i-1}, t)]}{\partial {\epsilon_i}^2}. \quad (7.5)$$

Setting $\partial f(\epsilon_i|\epsilon_{i-1}, t)/\partial t = 0$ in (7.5) and integrating twice, we obtain that the stationary distribution $f(\epsilon_i|\epsilon_{i-1})$ is a gamma distribution:

$$f(\epsilon_i|\epsilon_{i-1}) = \frac{(\beta_i/\epsilon_{i-1})^{\beta_i}}{\Gamma(\beta_i)} \epsilon_i^{\beta_i - 1} \exp\left(-\frac{\beta_i \epsilon_i}{\epsilon_{i-1}}\right), \quad (7.6)$$

where $\beta_i = 2\gamma_i/\kappa_i^2$.

Once we have the conditional distributions $f(\epsilon_i|\epsilon_{i-1})$, we can obtain the density $f(\epsilon_N)$ at the shortest scale by integrating over the intermediate scales:

$$f(\epsilon_N) = \int \prod_{i=1}^{N-1} d\epsilon_i\, f(\epsilon_N|\epsilon_{N-1}) \ldots f(\epsilon_1|\epsilon_0). \quad (7.7)$$

Remarkably, this multiple integral has an analytical representation in terms of a Meijer G-function [24], which is expressed as

$$f(\epsilon_N) = \frac{\omega}{\epsilon_0 \Gamma(\boldsymbol{\beta})} G_{0,N}^{N,0}\left(\begin{matrix} - \\ \boldsymbol{\beta} - \mathbf{1} \end{matrix} \middle| \frac{\omega \epsilon_N}{\epsilon_0}\right), \quad (7.8)$$

where $\omega = \prod_{j=1}^{N} \beta_j$ and we have introduced the vector notation $\boldsymbol{\beta} \equiv (\beta_1, \ldots, \beta_N)$ and $\Gamma(\mathbf{a}) \equiv \prod_{j=1}^{N} \Gamma(a_j)$. In our notation, a dash indicates the absence of parameters in the top row of the Meijer G-function, which comes from the fact that the first lower index is null.

Finally, by substituting Eqs. (7.1) and (7.8) into the compound integral (7.2), we obtain

$$P(x_N) = \frac{1}{\sqrt{2\pi}} \int_0^{\infty} \exp\left[-\frac{x_N^2}{2\epsilon_N}\right] \epsilon_N^{-1/2} f(\epsilon_N) d\epsilon_N. \tag{7.9}$$

Using the properties of the Meijer G-function, one can perform the integration above explicitly, which yields

$$P(x) = \frac{\omega^{1/2}}{\sqrt{2\pi\epsilon_0}\Gamma(\boldsymbol{\beta})} G_{0,N+1}^{N+1,0}\left(\begin{matrix} - \\ \boldsymbol{\beta} - \mathbf{1/2}, 0 \end{matrix} \middle| \frac{\omega x^2}{2\epsilon_0} \right), \tag{7.10}$$

where we omitted the subscript from x_N. It is worth mentioning that the large-x asymptotics of Eq. (7.10) has the form of a heavy-tailed modified stretched exponential [23],

$$P(x) \sim x^{2\theta} \exp\left\{-(N+1)[\omega(x^2/2\epsilon_0]^{1/(N+1)}\right\}, \quad |x| \to \infty, \tag{7.11}$$

where $\theta = (\sum_{i=1}^{N} \beta_i - N)/(N+1)$, which shows clear deviations from Gaussian tails.

We conclude this subsection by remarking that when the empirical data contains some internal structure indicating the presence of clusters of statistically independent samples, it may be necessary to employ a more general family of distributions given by a discrete statistical mixture,

$$P(x) = \sum_{j=1}^{n} p_j P_j(x), \tag{7.12}$$

where the statistical weights p_j are properly normalized, so that $\sum_{j=1}^{n} p_j = 1$, and each $P_j(x)$ is obtained from Eq. (7.10) for the same number N of hierarchy scales.

7.2.2 Entropy Formulation

In the dynamical formulation of H-theory, one tries to model the fluctuations in the environment of the system using a set of coupled stochastic differential equations. The theory can in principle provide a complete statistical description of the dynamics, which includes time dependent correlations functions of arbitrary orders. However, in many applications we are interested only in the stationary distribution of a relevant macroscopic variable. In this case, one can bypass the need to include much of the specifics of the stochastic dynamics, such as its memory, and take a thermodynamic-like approach in which the background distribution is derived from a maximum entropy principle. The main idea in the entropy approach is to write the Boltzmann-Shannon entropy of the system in terms of the local equilibrium distribution of states of the small system under observation and the distributions of the background variables, so that the system is treated as being effectively composed of a small system coupled to a set of nested reservoirs of increasingly larger size. The entropy formulation of H-theory [25] thus provides a rather general framework to describe the statistics of fluctuations in complex systems with multiple time and length scales.

We start by writing the Fokker–Planck equation associated with Eq. (7.3), which is given by

$$\frac{\partial P}{\partial t} = -\sum_{j=1}^{N} \frac{\partial J_j}{\partial \epsilon_j}, \tag{7.13}$$

where

$$J_j = -\gamma_j(\epsilon_j - \epsilon_{j-1})P - \frac{\kappa_j^2}{2}\frac{\partial}{\partial \epsilon_j}(\epsilon_j \epsilon_{j-1} P). \tag{7.14}$$

To be consistent with (7.7), we assume widely separated time scales $\gamma_N \gg \gamma_{N-1} \gg \ldots \gg \gamma_1$ and $\kappa_N \gg \kappa_{N-1} \gg \ldots \gg \kappa_1$, so that the condition of microscopic reversibility $J_j = 0$ applies for the equilibrium distribution. We thus have

$$\frac{\partial}{\partial \epsilon_j}(\epsilon_j \epsilon_{j-1} P_{eq}) = -\beta_j(\epsilon_j - \epsilon_{j-1})P_{eq}, \tag{7.15}$$

where $\beta_j = 2\gamma_j/\kappa_j^2$ are free parameters. Note that Eq. (7.15) implies that the stationary distribution $P_{eq}(\boldsymbol{\epsilon})$, where $\boldsymbol{\epsilon} = (\epsilon_1, \ldots, \epsilon_N)$ can be written in the factorized form shown in (7.7).

We may now consider the dynamical entropy, which is defined as

$$S(t) = -\int P(\boldsymbol{\epsilon}, t) \ln P(\boldsymbol{\epsilon}, t) d\boldsymbol{\epsilon}, \tag{7.16}$$

where we use the shorthand notation $d\boldsymbol{\epsilon} = \prod_{j=1}^{N} d\epsilon_j$. We may use the existence of $P_{eq}(\boldsymbol{\epsilon})$ to write Eq. (7.16) as $S(t) = S_i(t) + S_e(t)$, where

$$S_i(t) = -\int d\boldsymbol{\epsilon}\, P(\boldsymbol{\epsilon}, t) \ln[P(\boldsymbol{\epsilon}, t)/P_{eq}(\boldsymbol{\epsilon})] \tag{7.17}$$

and

$$S_e(t) = -\int d\epsilon\, P(\epsilon, t) \ln P_{eq}(\epsilon). \quad (7.18)$$

Here the two terms of entropy have their changes, $dS_i(t)$ and $dS_e(t)$, related to irreversible entropy production and reversible entropy flow, respectively [26, 27]. Taking the derivative of $S(t)$ with respect to time we get

$$\dot{S}_i(t) = -\sum_{j=1}^{N} \int d\epsilon J_j \frac{\partial}{\partial \epsilon_j} \ln[P(\epsilon, t)/P_{eq}(\epsilon)] \quad (7.19)$$

and

$$\dot{S}_e(t) = -\frac{d}{dt} \langle \ln P_{eq}(\epsilon) \rangle_t . \quad (7.20)$$

The sign of $\dot{S}_e(t)$ depends on the initial condition $P_0(\epsilon) \equiv P(\epsilon, 0)$. In contradistinction, on using (7.14) and (7.15) we see that

$$\dot{S}_i(t) = \sum_{j=1}^{N} \int d\epsilon \frac{2J_j^2}{\kappa_j^2 \epsilon_j \epsilon_{j-1} P(\epsilon, t)} \geq 0. \quad (7.21)$$

We thus conclude that the stationary solution $P_{eq}(\epsilon)$ is a dynamical maximum of $S_i(t)$. This result is consistent with a more general analysis of the entropy evolution of stochastic dynamical systems [28]. A Fokker–Planck approach to systems governed by a hierarchical dynamics has also been considered in the literature [29]. To proceed further we use $p(\boldsymbol{\beta}) = \prod_{j=1}^{N} f\left(\beta_j | \beta_{j-1}\right)$ to write

$$\ln P_{eq}(\epsilon) = \sum_j \ln f(\epsilon_j | \epsilon_{j-1}) \quad (7.22)$$

and thus

$$S_i(t) = -\int d\epsilon\, P(\epsilon, t) \left[\ln P(\epsilon, t) - \sum_j \ln f(\epsilon_j | \epsilon_{j-1}) \right]. \quad (7.23)$$

We may now extend S_i to a Lagrange functional $S_i[P]$ and surmise that it reaches a conditional maximum at the stationary solution, $P = P_{eq}$, with the averages $\langle \ln f(\epsilon_j | \epsilon_{j-1}) \rangle$ being constraints. This is the basic principle of the maximum entropy approach [25].

7.2.3 CLT Approach

The dynamical stochastic approach to H-theory, described in the previous subsections, allows for a clear connection with fluid turbulence, whilst the maximum entropy formulation shows that some of the specific details of the background dynamics, such as its memory and time-dependent correlation functions, are irrelevant for the description of $P(x)$ in the stationary regime. However, neither provide a direct interpretation of the discrete parameter N, i.e., the number of levels or relevant scales in the model hierarchy. In other words, it is not always clear which structure should play the role of the turbulent eddies in the complex dynamical system described by the H-theory. In order to address this issue, we introduce below a discrete model that makes direct contact with the central limit theorem (CLT) and provides a simple interpretation for the number N of hierarchical levels present in H-theory [30]. A related approach, albeit restricted to the cases $N = 0$ and $N = 1$ of a two-dimensional random walk, can be found in Ref. [31].

We start by defining X as a random variable, with $\{X_i\}_{i=1,\ldots,M}$ denoting a set of M independent realizations of X. We also define the random variable Z via $Z_i \equiv X_i - \langle X \rangle$, and let $Y_M = \sum_{i=1}^{M} Z_i$. The characteristic functions of Y_M and Z are related as follows:

$$\Phi_{Y_M}(k; M) = \langle e^{ikY_M} \rangle = \langle e^{ikZ} \rangle^M = [\Phi_Z(k; M)]^M . \tag{7.24}$$

In the standard form of the CLT, we consider M to be a large number, so that

$$\Phi_{Y_M}(k; M) = \left[1 - \frac{k^2}{2M^2}\sigma_X^2 + \cdots\right]^M \simeq \exp\left(-\frac{k^2\sigma_X^2}{2M}\right), \tag{7.25}$$

where σ_X is the root-mean-square deviation of X. This characteristic function implies that Y_M is also Gaussian distributed, as expected from the CLT.

Let us consider a discrete random variable M with probability P_M. We define a weighted averaged characteristic function,

$$\chi_Y(k) = \sum_{M=0}^{\infty} P_M \Phi_{Y_M}(k; M). \tag{7.26}$$

We proceed by selecting P_M from particular stationary solutions of well-known discrete stochastic processes. From these results, we will be able to make contact with the H-theory and the associated interpretation of the number of hierarchy scales N.

(i) Poisson process ($N = 0$)
The Poisson process is described by the master equation [32],

$$\dot{P}_M(t) = \gamma[P_{M-1}(t) - P_M(t)], \tag{7.27}$$

which represents an interaction-free evolution in discrete space, with $\dot{P}_M(t) \equiv dP_M/dt$ and γ a positive parameter. The solution is

$$P_M = \frac{\bar{M}^M}{M!}e^{-\bar{M}}, \tag{7.28}$$

where $\bar{M} = \gamma t$ and the characteristic function is given by $\chi_Y(k) = \exp\left[\bar{M}\left(\Phi_Z(k) - 1\right)\right]$. If we rescale $Z \to Z/\sqrt{\bar{M}}$ and take $\bar{M} \to \infty$, we recover the expected result of the CLT, $\chi_Y(k) \to \exp(-k^2/2\sigma_X^2)$.

(ii) Pauli process ($N = 1$)
The master equation of the Pauli process is defined as

$$\dot{P}_M(t) = \zeta_{M+1}P_{M+1}(t) + \gamma_{M-1}P_{M-1}(t) - (\zeta_M + \gamma_M)P_M(t), \tag{7.29}$$

where $\zeta_M = aM$ and $\gamma_M = b(M+1)$, with $a > 0$ and $b > 0$. The stationary solution reads $P_M = (1-u)u^M$, where $u = b/a = \bar{M}/(1+\bar{M})$ and $\bar{M} = \sum_{M=0}^{\infty} MP_M$ and the characteristic function is

$$\chi_Y(k) = \left[1 - \bar{M}\left(\Phi_Z(k) - 1\right)\right]^{-1}. \tag{7.30}$$

Now, if we rescale $Z \to Z/\sqrt{\bar{M}}$ and take $\bar{M} \to \infty$, we obtain

$$\chi_Y(k) = \left(1 + \frac{k^2}{2\sigma_X^2}\right)^{-1}. \tag{7.31}$$

The corresponding distribution for the variable Y is

$$P(y) = \frac{\sigma_X}{\sqrt{2\pi}}G_{0,2}^{2,0}\left(\begin{matrix} - \\ 1/2,\, 0 \end{matrix}\middle| \frac{\sigma_X^2 y^2}{2}\right), \tag{7.32}$$

which is a particular case of Eq. (7.10).

(iii) Negative binomial process ($N = 1$)
The negative binomial process is a generalization of the Pauli process that allows for the presence of bunches in the evolution

of P_M [32]. In this case, the stationary solution is

$$P_M = \binom{M+\nu-1}{M} \frac{a^M}{(1+a)^{M+\nu}}, \tag{7.33}$$

where $a = \bar{M}/\nu$ and the parameter ν is the number of degrees of freedom. Notice that the binomial coefficient counts the number of partitions of M among ν states. The characteristic function reads

$$\chi_Y(k) = \left[1 - \frac{\bar{M}}{\nu}(\Phi_Z(k) - 1)\right]^{-\nu}. \tag{7.34}$$

If we now rescale $Z \to Z/\sqrt{\bar{M}}$ and take $\bar{M} \to \infty$, we obtain

$$\chi_Y(k) = \left(1 + \frac{k^2}{2\sigma_X^2}\right)^{-\nu}. \tag{7.35}$$

The corresponding distribution for the variable Y is

$$P(y) = \frac{\nu^{1/2}\sigma_X}{\sqrt{2\pi}\Gamma(\nu)} G_{0,2}^{2,0}\left(\begin{matrix} - \\ \nu - 1/2, 0 \end{matrix} \middle| \frac{\nu\sigma_X^2 y^2}{2}\right), \tag{7.36}$$

which is also a particular case of Eq. (7.10). Notice that for $\nu = 1$ we recover the results of the Pauli process.

We are now in position to state the general result for arbitrary N. Consider a stochastic process with stationary distribution given by

$$P_M^{(N)} = \frac{1}{M!\Gamma(\boldsymbol{\nu})} G_{N,1}^{1,N}\left(\begin{matrix} \mathbf{1} - \boldsymbol{\nu} \\ M \end{matrix} \middle| a\right), \tag{7.37}$$

where $\bar{M} = a\bar{\nu}$ and $\bar{\nu} = \prod_j \nu_j$. The integer parameter N gives the order of the transitions induced by the nonvanishing coupling between $P_{M\pm N}^{(N)}(t)$ and $P_M^{(N)}(t)$ in the master equation. Observe that for $N = 0$ (no correlations) and $N = 1$ (dipole transitions) we recover the Poisson and negative binomial distributions, respectively. The corresponding distribution for the variable Y can be expressed as

$$P_N(y) = \frac{\bar{\nu}^{1/2}\sigma_X}{\sqrt{2\pi}\Gamma(\boldsymbol{\nu})} G_{0,N+1}^{N+1,0}\left(\begin{matrix} - \\ \boldsymbol{\nu} - \mathbf{1/2}, 0 \end{matrix} \middle| \frac{\bar{\nu}\sigma_X^2 y^2}{2}\right), \tag{7.38}$$

which reproduces the exact stationary solution (7.10) of H-theory. We can now interpret the parameter N in the H-theory. If we consider Y_M to be a sum of M events that contribute to the increments of the observed time series, then N is essentially the range of the coupling between background variables with different time scales. In the next section, we shall use this argument in the context of turbulence in RFL.

7.3 Turbulence-Like Properties in a RFL

In a conventional laser the basic components are an optical cavity and a gain medium. Coherent light can be generated by optically pumping the gain medium, where spontaneous emission occurs. The optical cavity, normally formed by mirrors, defines spatial modes and provides radiation feedback to the gain medium. Coherent light is then amplified, due to stimulated emission in the optical modes. Above a certain pumping threshold, the gain overcomes the losses and laser radiation is produced. By contrast, a RL system provides laser-like radiation without relying on fixed mirrors to provide optical feedback. This is achieved by exploring a multiple scattering mechanism which extends to the optical path of photons created by spontaneous emission inside the gain medium. This creates an amplification of the emitted light due to stimulated emission, in a process called amplified spontaneous emission. The radiation emitted by a RL is predominantly multimode, which implies that its coherence properties are poorer than that of a conventional laser. In imaging, the low spatial coherence of RLs produces speckle patterns with lower visibility, which can be used to produce smoother images [33]. The low temporal coherence makes RLs attractive for applications in optical coherence tomography, since for this imaging technique the shorter the light source coherence length, the better [34].

Here we employ the statistical approach (H-theory) described in Section 7.2 to explain the non-Gaussian behavior of the time series of intensity increments in Er-RFL, in which disorder is present in the form of customized random Bragg grating scatterers. The basic idea connecting H-theory with RL emission builds on the theory of exponential amplification of diffusing spontaneous photons [35]. To be more specific, let $I_i(\ell_i, \lambda)$ be the emission intensity of a path i that started with a photon of wavelength λ with a trajectory of length ℓ_i, then

$$I_i(\ell_i, \lambda) = I_0 e^{\ell_i/L_G}, \tag{7.39}$$

where L_G the gain length of the active medium, defined as the mean distance traveled for the first amplification of the photon and I_0 is an intensity threshold. The length ℓ_i of the photon trajectory is a random variable and thus the total emission intensity, which is given by

$$I(\lambda) = \sum_{i=1}^{M} I_i(\ell_i, \lambda), \tag{7.40}$$

is a sum of a random number of random variables. This is exactly the starting point of the CLT approach to H-theory, described in Subsection 7.2.3.

In the quantitative analysis, we employed the same RFL system and dataset described in Ref. [20], where complete experimental details can be found. We remark that an exceptionally large number ($\sim 10^3$) of random Bragg gratings were inscribed in the multimode Er-doped fiber core to strengthen the degree of quenched disorder. The system was pumped by a CW semiconductor laser. An extensive number (N_s = 150,000) of emission spectra were collected for a value of the excitation power above the RL threshold (P/P_{th} = 2.92). A very long series of output intensities $\{I_\alpha(k)\}$, $\alpha = 1, \ldots, N_s$, was thus generated for each one of the 512 wavelengths indexed by k in the interval [1489.0 nm, 1591.4 nm] around the peak emission at 1540.0 nm. We notice that α can be also related to the time variable $t = \alpha t_0$, where t_0 = 100 ms is the integration time between two consecutive spectra acquisitions.

In Figs. 7.1(a–c), we show representative samples of 500 emission spectra for each excitation power, respectively in the regimes below ($P/P_{\text{th}} = 0.72$), near ($P/P_{\text{th}} = 0.99$), and above threshold ($P/P_{\text{th}} = 2.92$), where the power threshold, P_{th} = 16.30 mW, was measured from a FWHM analysis. We determined the maximum intensity in each spectrum of the whole data set and thus obtained a long time series of fluctuating intensity values $I(t)$, with the dimensionless time index $t = 1, 2, \ldots,$ 150,000, which is illustrated in Figs. 7.1(d–f) for the respective values of P/P_{th}. It is clear from these plots that there is a dramatic change in the fluctuation pattern

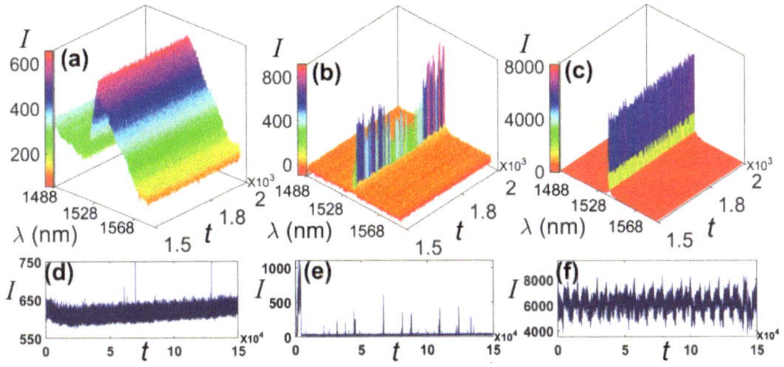

Figure 7.1 Emission spectra and time series of maximum intensities in Er-RFL. (a)-(c) Representative samples showing 500 successive emission spectra at each excitation power P (normalized by the threshold value P_{th}) below [(a) $P/P_{\mathrm{th}} = 0.72$], near [(b) $P/P_{\mathrm{th}} = 0.99$], and above [(c) $P/P_{\mathrm{th}} = 2.92$] the threshold of the Er-RFL. Intensities I are displayed in arbitrary units as a function of the wavelength λ. The portrayed spectra were collected at times $t\tau$, with integration time window $\tau = 100$ ms and dimensionless index $t = 1501, 1502, \ldots, 2000$. (d)–(f) For each one of the 150,000 spectra in the whole data set, the maximum intensity was recorded, resulting in the long time series $I(t)$, with $t = 1, 2, \ldots, 150{,}000$, which is shown in (d), (e), and (f) for the respective values of P/P_{th}. (Reproduced with permission from Ref. [20].)

of $I(t)$ as the excitation power crosses the threshold. In fact, we shall see below that this transition actually marks the onset of turbulent emission in the Er-RFL system.

In analogy to fluid turbulence, where the relevant statistical quantities are velocity increments between two points (rather than the velocities themselves), we shall analyze here the intensity increments, $\delta I(t) \equiv I(t+1) - I(t)$, between successive optical spectra. More specifically, we define the signal associated with the intensity fluctuations as the zero average and unit variance stochastic process given by $x(t) \equiv \delta I(t)/\sqrt{\mathrm{var}(\delta I(t))}$, where $\mathrm{var}(\delta I(t))$ denotes the variance of the time series of intensity increments. If nonlinearities are not relevant, as in the prelasing regime, then the intensity increments are statistically independent and the probability distribution $P(x)$ is a Gaussian.

On the other hand, as the excitation power is increased beyond the threshold, we show below that dynamical nonlinearities give rise to turbulent emission. This implies that the Gaussian form of the signal distribution remains valid only at a local level, acquiring a slowly fluctuating variance ε along the time series. In this case, as described in Section 7.2, we can still write the local distribution of x as a conditional Gaussian $P(x|\varepsilon) = \exp\left(-x^2/\,2\varepsilon\right)/\sqrt{2\pi\varepsilon}$, where the parameter ε characterizes the local equilibrium.

The non-Gaussian form of $P(x)$ can thus be obtained by compounding the local Gaussian with the background distribution $f(\varepsilon)$ of the variance ε, as given in Eq. (7.2). Furthermore, adopting the stochastic model for $f(\varepsilon)$ presented in Section 7.2, we then obtain that the probability density $P(x)$ of the intensity increments x can be given in terms of the hierarchical distributions shown in (7.10), as will be discussed in more detail below.

As a first quantitative characterization of the experimental time series $I(t)$ of maximum intensities, shown in Figs. 7.1(d–f), we calculated the power spectral density,

$$S(k) = \left|\sum_{n=1}^{L} I_n \exp\left(\frac{-2\pi i(n-1)(k-1)}{L}\right)\right|^2, \tag{7.41}$$

by subdividing the time series into 550 windows of size $L = 256$, evaluating $S(k)$ for each window and then performing the average of $S(k)$ over all windows. Here, I_n denotes the n-th value of the intensity inside the window and k/L is the dimensionless frequency. In Ref. [20] we present log-log plots of $S(k)$ for the three mentioned values of the excitation power: (a) $P/P_{\text{th}} = 0.72$, (b) $P/P_{\text{th}} = 0.99$, and (c) $P/P_{\text{th}} = 2.92$, corresponding, respectively, to the regimes below, near, and above the threshold. The lines observed [20] are nicely fitted to the power-law behavior $S(k) \sim k^{-\alpha}$. The white noise ($\alpha = 0$) below threshold is consistent with the statistically independent nonturbulent intensity fluctuations, displaying Gaussian distributions for both intensities and intensity increments.

In contrast, the nontrivial double power-law behavior noticed both near and above threshold suggests the existence of turbulence in the intensity fluctuations dynamics, which is confirmed below. Such behavior of $S(k)$ is also consistent with the double cascade phenomenon observed in the wave turbulence scenario [36].

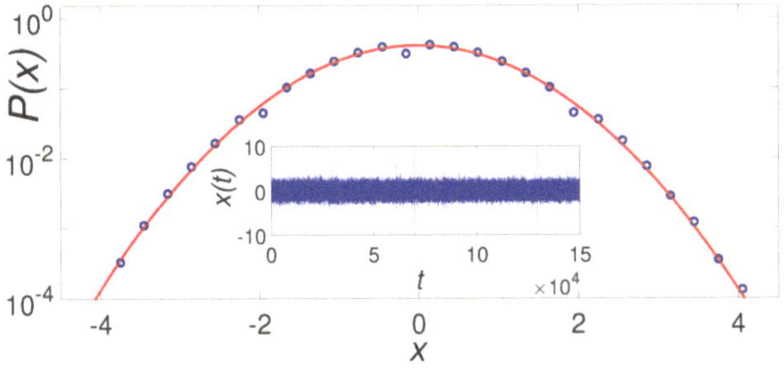

Figure 7.2 Gaussian distribution of intensity increments and nonturbulent prelasing behavior in Er-RFL. Semi-log plot of the distribution $P(x)$ of experimental intensity increments x in the nonturbulent prelasing regime at the excitation power (normalized by the threshold value) $P/P_{\text{th}} = 0.72$ (blue circles) and best fit to a Gaussian function (red line). The intensity increments are defined as the zero average and unit variance stochastic process given by $x(t) \equiv \delta I(t)/\sqrt{\text{var}(\delta I(t))}$, where $\delta I(t) \equiv I(t+1) - I(t)$ denotes the fluctuations of maximum intensities in 150,000 successive optical spectra and var$(\delta I(t))$ is the variance of the time series of intensity increments. The inset displays the corresponding time series of $x(t)$ as function of the dimensionless time index $t = 1, 2, \ldots, 150{,}000$. (Reproduced with permission from Ref. [20].)

In Fig. 7.2, we display the experimental distribution of intensity increments below threshold, at $P/P_{\text{th}} = 0.72$. The good Gaussian fit corroborates the results for the prelasing regime in the analysis of $S(k)$. The variance ε does not fluctuate, corresponding to the density $f(\varepsilon)$ given by a Dirac delta function in the compound integral, Eq. (7.9). Consequently, a multiscale turbulent cascade is absent below threshold ($N = 0$ in the model), which implies that the intensity fluctuations in Er-RFL produces intensity increments that are statistically independent.

The scenario is drastically modified as the excitation power is increased near and above the threshold. Indeed, the excellent fit to the distribution of intensity increments observed in Fig. 7.3(a) at $P/P_{\text{th}} = 0.99$ is described by the statistical mixture,

$$P(x) = pP_N(x; \varepsilon_0, \beta) + (1-p)P_N(x; \varepsilon_0', \beta'), \tag{7.42}$$

where $P_N(x; \varepsilon_0, \beta)$ is given by Eq. (7.10), with $\beta_j = \beta$ for all j. The values of the parameters are $N = 1$ (K-distribution), $p = 0.94$, $\beta = 4.21$, $\epsilon_0 = 0.36$, $\beta' = 1.2$, $\epsilon_0' = 12.0$. In this case, the statistical

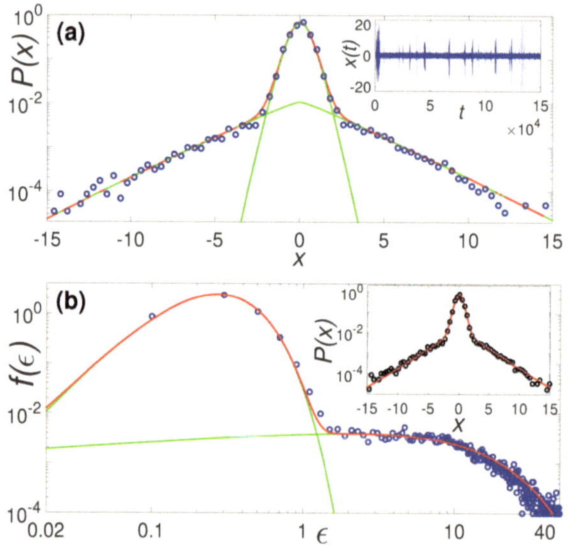

Figure 7.3 Statistical mixture of intensity increments and turbulent behavior near the threshold in Er-RFL. (a) Semi-log plot of the distribution $P(x)$ of experimental intensity increments x (blue circles) near the threshold, at the excitation power (normalized by the threshold value) $P/P_{\mathrm{th}} = 0.99$, and the prediction from the hierarchical model (7.3) for the statistical mixture (7.42) (red line). The individual components of the mixture, given by two K-distributions with $N = 1$, are shown by green lines. The inset shows the corresponding time series of intensity increments. A single time scale ($N = 1$) in the dynamics of the background variable characterizes the turbulent behavior of the intensity fluctuations dynamics in Er-RFL near the threshold. (b) Log-log plot of the density function $f(\epsilon)$ of the experimental variance series $\epsilon(t)$ (blue circles) and the hierarchical model prediction for the statistical mixture (7.43) (red line), with same parameters as in (a). The individual components of the statistical mixture are shown by green lines. The inset displays the compounding of $\epsilon(t)$ with a Gaussian function (red line) and the experimental distribution (black circles). (Reproduced with permission from Ref. [20].)

model predicts a single time scale ($N = 1$) influencing the dynamics of the background variable (that is, the fluctuating variance of intensity increments), which in turn characterizes the turbulent behavior of the fluctuation dynamics of the output intensity in Er-RFL near the threshold.

In order to verify the compound hypothesis expressed in Eqs. (7.1) and (7.6), we implemented a procedure to compute a subsidiary time series of variance estimators $\epsilon(t)$. To this end, the time series $x(t)$ of intensity increments was subdivided into overlapping intervals of size M, and for each such interval we computed the variance estimator, $\epsilon(t) = \frac{1}{M}\sum_{j=1}^{M}[x(t-j) - \bar{x}(t)]^2$, where $\bar{x}(t) = \frac{1}{M}\sum_{j=1}^{M} x(t-j)$ and $t = M, \ldots, 150000$, thus generating a new time series. Next, we numerically compounded the distribution of $\epsilon(t)$ with a Gaussian function, as suggested by Eq. (7.9), for various M, and selected the value of M for which the corresponding superposition integral best fitted the distribution of the original time series. Excellent agreement for $P(x)$ was found using $M = 15$, as seen in the inset of Fig. 7.3(b). The optimal value of M can be interpreted as an estimation of the large time scale associated with fluctuations in the background variable ε. In the main plot of Fig. 7.3(b), we show the good agreement between the distribution related to the $\epsilon(t)$-series generated using the optimal window size M and the background density $f(\varepsilon_N)$ with statistical mixture,

$$f(\varepsilon_N) = pf_N(\varepsilon_N; \varepsilon_0, \beta) + (1-p)f_N(\varepsilon_N; \varepsilon_0', \beta'), \tag{7.43}$$

where $f_N(\varepsilon_N; \varepsilon_0, \beta)$ is given by Eq. (7.8), for the same parameters as in Fig. 7.3(a), as expected from the consistency with the joint fit of statistical mixtures. The individual components of the mixture are shown by green lines in both Figs. 7.3(a) and 7.3(b).

Figure 7.4(a) shows the distribution of intensity increments well above the threshold, at $P/P_{\text{th}} = 2.92$. The red line displays the fit to a statistical mixture in the form of Eq. (7.42), with parameters $N = 6$, $p = 0.30$, $\beta = 8.3$, $\epsilon_0 = 0.19$, $\beta' = 6.5$, $\epsilon_0' = 1.3$. The background distribution is shown in Fig. 7.4(b), along with the fit (red line) to the statistical mixture as in Eq. (7.43), with the same parameters. Excellent agreement is found in both fits. The superposition of the distribution of the variance estimator ($M = 22$) with a Gaussian distribution is depicted by the red line in the inset of Fig. 7.4(b). The individual components of the mixture are shown by green lines in both Figs. 7.4(a) and 7.4(b). We observe that the changes in the parameters β, β', ϵ_0, ϵ_0', and p at $P/P_{\text{th}} = 2.92$, in comparison with the respective values near the threshold,

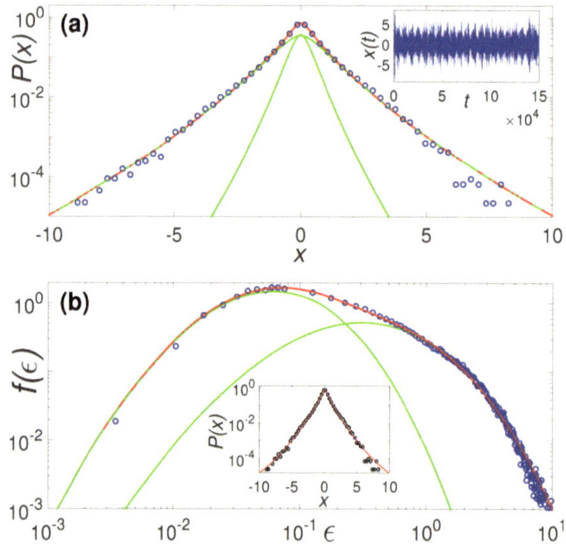

Figure 7.4 Statistical mixture and multiscale turbulent behavior above the threshold in Er-RFL. (a) Semi-log plot of the distribution $P(x)$ of experimental intensity increments x (blue circles) in the regime well above threshold, at the excitation power (normalized by the threshold value) $P/P_{\rm th} = 2.92$, and the prediction from the hierarchical model (7.3) for the statistical mixture (7.42) (red line). The individual components of the mixture, given by two Meijer G-distributions with $N = 6$, are shown by green lines. The expected rise in the number of relevant time scales ($N = 6$) upon increasing the excitation power is consistent with the multiscale turbulent cascade behavior of the intensity fluctuations dynamics in Er-RFL above threshold. The inset shows the corresponding time series of intensity increments. (b) Log-log plot of the density function $f(\epsilon)$ of the experimental variance series $\epsilon(t)$ (blue circles) and the hierarchical model prediction for the statistical mixture (7.43) (red line), with same parameters as in (a). The individual components of the statistical mixture are shown by green lines. The inset displays the compounding of $\epsilon(t)$ with a Gaussian function (red line) and the experimental distribution (black circles). (Reproduced with permission from Ref. [20].)

accommodate well the changes in the shapes of the experimental distributions for both $x(t)$ and $\varepsilon(t)$ caused by an enhancement of the intermittency effect. For example, the decrease in the weighting parameter p (from $p = 0.94$ at $P/P_{\rm th} = 0.99$ to $p = 0.30$ at $P/P_{\rm th} = 2.92$) indicates an increase in the statistical relevance of the underlying structure associated with the higher mean variance ε'_0, whose weight is $(1 - p)$, which is reasonable since fluctuations become stronger above the threshold. Similarly, the observed value of N is also consistent with the expected rise in the number of relevant time scales upon increasing the excitation power.

In this respect, we also notice that the emergence of K-distributions in weak-scattering by continuous media, such as sea clutter, has been attributed [37] to the modulation of small-scale fluctuations by more slowly changing large-scale structures, a scenario that is consistent with our hierarchical model with a single time scale for the background ($N = 1$). In Er-RFL near the threshold, a similar mechanism may develop whereby large structures consisting of groups of correlated scatterers may (intermittently) form, thus yielding $N = 1$, as verified above. Beyond the threshold, stronger nonlinearities may lead to multiscale dynamics where small-scale intensity fluctuations are modulated by larger scales which, in turn, are modulated by even larger scales, and so on, up to the largest scale in the system. This cascade-like behavior thus implies the existence of a number $N > 1$ of characteristic scales—a view that is precisely captured by the hierarchical model (Eq. 7.3)— with N expected to grow as the excitation power increases, provided there is no saturation of the gain or the nonlinearity. We remark, however, that although the existence of a turbulence hierarchy in the Er-RFL system has been clearly identified from a statistical analysis of a large set of emission spectra, more studies are necessary to develop a comprehensive understanding of the underlying physical mechanisms.

From the analysis above, we observe that in the regimes near and above the threshold, where the roles of disorder and nonlinearity are mostly evidenced in the Er-RFL system, the theoretical description is possible only through a mixture of statistics that arises from a hierarchical stochastic model for the multiscale intensity fluctuations, which is strictly related to the emergence of Kolmogorov's turbulence behavior in the distribution of intensity increments.

In this sense, the extensive size of the experimental data set was proved essential to unveil the turbulent emission behavior in Er-RFL. Indeed, from the discussion above we infer that a significant analysis of the variance fluctuations in the compound integral should require at least $\sim 10^5$ emission spectra.

We should however point out that discrete statistical mixtures are usually associated with a partition of the data into statistically independent subsets. In [38], for instance, the probability density function of the dissipation rate of kinetic energy in oceanic turbulence has been found to be bimodal and well described by a mixture of two log-normal distributions. In this case, the authors speculate that the partition of the data was due to a combination of an active mode and a quiescent one. While we did not attempt to separate our experimental data according to some identified mechanism responsible for the statistical mixtures observed both near and above the threshold, we can infer from general grounds that such mechanism could arise from a subtle combination of stimulated and spontaneous turbulent emissions in the presence of both nonlinearity and disorder.

7.3.1 Intermittency in RFL Turbulent-Like Emission

In fluid turbulence, intermittency is usually explained as being caused by fluctuations in the energy transfer rates or energy fluxes between adjacent scales, and its detailed description is one of the goals of current research. A universal feature of intermittence is that its effect vanishes monotonically as we move from small to large scales. We shall verify this universal property in the RFL photonic turbulent-like state presented below.

We start the analysis by generating a matrix from the large dataset of intensity measurements $\{I_\alpha(k)\}$ at the excitation power $P/P_{\text{th}} = 2.92$, for $\alpha = 1, \ldots, N_s$ (= 150,000) output spectra replicas (matrix rows) and 512 wavelengths indexed by k in the interval [1489.0 nm, 1591.4 nm] (matrix columns). We then perform a subtraction of replicas (rows) separated by τ units of time. In other words, we work with one intensity dataset for each wavelength indexed by k (i.e., for each column of the matrix), and also calculate the average over replicas (rows) for each k. In this procedure, each new matrix generated as above contains (150,000 $-$ τ) rows and 512 columns. We then perform the calculation of the distribution of the variable

$$x_{\alpha\tau}(k) = \frac{\delta I_{\alpha\tau}(k)}{[\sum_{\alpha} \delta I^2_{\alpha\tau}(k)]^{1/2}}, \tag{7.44}$$

for each given choice of τ measured at the wavelength 1539.8 nm around the peak emission (value of k corresponding to the matrix column 254).

Plots of such distributions are shown in Fig. 7.5 [(a,b) $P(x_1)$; (c,d) $P(x_2)$ and (e) $P(x_{5000})$]. In all cases, we notice an excellent agreement between the theoretical and experimental results for all values of τ. By applying the numerical procedure described above, we obtain a statistical mixture, defined as $P(x_\tau) = p_\tau P_1(x_\tau) + (1 - p_\tau)P_2(x_\tau)$, where $P_j(x_\tau)$ is given by (7.10) with fitting parameters N_τ, $\beta_\tau^{(j)}$ and $\epsilon_\tau^{(0,j)}$. For the separation time $\tau = 1$ between the RFL emission spectra, the distributions $P(x_1)$ [(a)] and $f(\epsilon_1)$ [(b)] with the best fit parameters $N_1 = 6$, $p_1 = 0.30$, $\beta_1^{(1)} = 8.53$, $\beta_1^{(2)} = 6.47$, $\epsilon_1^{(0,1)} = 0.16$ and $\epsilon_1^{(0,2)} = 1.36$. On the other hand, for $\tau = 2$, the plots of $P(x_2)$ [(c)] and $f(\epsilon_2)$ [(d)], obtained for $N_2 = 5$, $p_2 = 0.27$, $\beta_2^{(1)} = 5.73$, $\beta_2^{(2)} = 6.07$, $\epsilon_2^{(0,1)} = 0.18$ and $\epsilon_2^{(0,2)} = 1.31$. Remarkably, the results for a large separation in time ($\tau = 5000$) [(e)] display the Gaussian profile of $P(x_{5000})$ ($N = 0$), in analogy to the nonturbulent pattern observed from the velocity increments of rather distant points in a turbulent fluid.

The effect of intermittency (non-Gaussian tails in $P(x_\tau)$) is large for $\tau = 1$ and $\tau = 2$, and is essentially irrelevant for $\tau = 5000$ and $\tau = 0$ (no intensity increments). Thus, for $\tau = 5000$ and $\tau = 0$ the variable $x_{\alpha\tau}(k)$ is Gaussian distributed, with no need to calculate $P(x_\tau)$ from a superposition integral such as Eq. (7.9) (this situation corresponds to set $N = 0$ in the hierarchical model, in which $f(\epsilon)$ in (7.9) is trivially given by a Dirac δ-function). For $\tau = 1$ and $\tau = 2$ intermittency effects are important, with large deviations from the Gaussian behavior being clearly observed. Nevertheless, the Gaussian form of the distribution of normalized intensity increments $\{x_{\alpha\tau}(k)\}$ still remains valid at a local level, with a slowing fluctuating variance ϵ_τ, so that the compound, or superposition integral, actually applies. In this case, the complex dynamics (intermittency) of the photonic turbulent state is captured by the background distribution $f(\epsilon_\tau)$ and since it can be explicitly calculated from the hierarchical model in the form of a Meijer-G

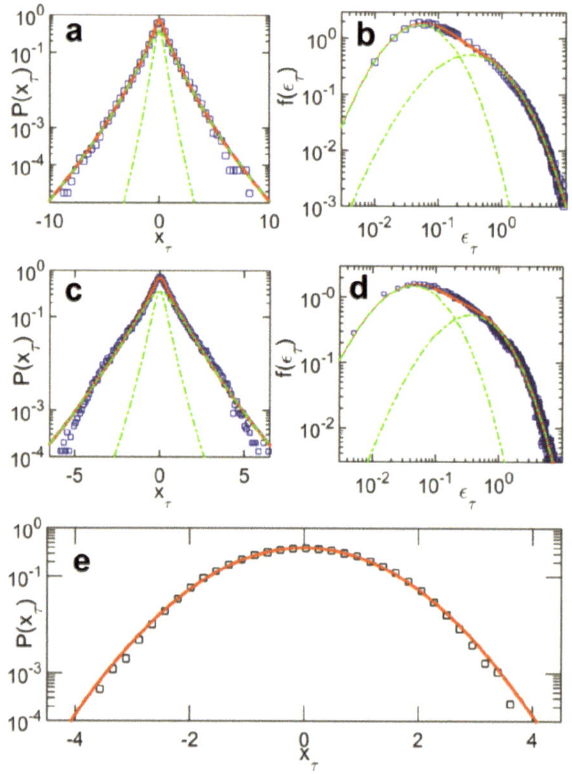

Figure 7.5 Distributions of intensity increments $P(x_\tau)$ and background density $f(\epsilon_\tau)$ for RFL emission spectra separated by τ units of time that are (a, b) first neighbors ($\tau = 1$), (c, d) second neighbors ($\tau = 2$), and (e) distant neighbors ($\tau = 5000$). The theoretical results (red lines) adjust quite well to the experimental data (squares), with the mixture components shown in green dashed lines. (Reproduced with permission from Ref. [39].)

function, Eq. (7.8), we have obtained $f(\epsilon_\tau)$, for $\tau = 1, 2$, directly from the experimental data and compared with the theoretical result. The excellent agreement that was found is a good evidence of the consistency of the photonic turbulent phase in the RFL. Finally, the values obtained for N_τ, which were $N_{5000} = 0$, $N_2 = 5$ and $N_1 = 6$, indicate that intermittency is accompanied by an increase in the number of relevant scales in the hierarchy of the photonic turbulent state.

Combining the experimental evidences obtained in our previous works on the hierarchical structure of the photonic turbulence regime in RFL with the observation of the intermittency crossover effect in the present work, we conclude that the validity of our hierarchical stochastic model appears to stand on solid ground. The next logical step is to move towards nonequilibrium thermodynamics and introduce a small external perturbation in the turbulent phase in order to observe the subsequent relaxation of the system, which could be compared with the predictions of H-theory. Measurements of the entropy production in the relaxation process would also be very interesting.

7.4 Coexistence of Turbulent and Spin-Glass Behaviors in a RFL

Turbulence and spin glass constitute two of the most challenging phenomena observed in complex systems. Recently, we demonstrated [39] the first evidence in any physical system of the coexistence of such behaviors. The system employed was a quasi-one-dimensional Er-RFL, which displays all essential ingredients underlying both behaviors, namely disorder, frustration and nonlinearity, as well as turbulent energy cascades and intermittent energy flux between fluctuation scales.

On the one hand, the theoretical framework suitable to describe the photonic turbulent behavior in the Er-RFL has been described in the previous sections. On the other hand, we refer the reader to Chapters 5, 6, and 8 of this book for a comprehensive discussion on both theoretical and experimental aspects of the photonic spin-glass phase.

The RFL was pumped by a CW semiconductor laser and a large number ($N_s = 150{,}000$) of emission spectra were collected for a value of the excitation power above the RL threshold ($P/P_{\text{th}} = 2.92$). A long series of measured intensities $\{I_\alpha(k)\}$, $\alpha = 1, \ldots, N_s$, was obtained for each one of the 512 wavelengths indexed by k in the interval [1489.0 nm, 1591.4 nm] around the peak emission at 1540.0 nm. The parameter α can be related to the time variable

$t = \alpha t_0$, where $t_0 = 100$ ms is the integration time between two consecutive spectra acquisitions. For $P/P_{\text{th}} = 2.92$ the RFL exhibits a photonic spin glass phase with replica symmetry breaking (RSB) [40, 41] and turbulent-like signature in the distribution of successive intensity increments [20].

In order to describe the coexistence of these complex behaviors, we shall define a Pearson correlation coefficient [42, 43] that measures the linear correlation between the fluctuation variables $y_{\alpha\tau}(k) \equiv I_{\alpha+\tau}(k) - cI_\alpha(k)$, where τ is the time difference (in units of t_0) between the spectra, see below.

However, before going into much details of this analysis, we would like to introduce the concept of the Pearson correlation coefficient [42, 43]. Consider two variables x and y with sets of values, $\{x_i\}$ and $\{y_i\}$, for $i = 1, 2, \ldots, N$. The Pearson correlation coefficient C_{xy} gives a measure of the normalized linear covariance between these two sets in the form

$$C_{xy} = \frac{\sum_i (x_i - \bar{x})(y_i - \bar{y})}{\sqrt{\sum_i (x_i - \bar{x})^2}\sqrt{\sum_i (y_i - \bar{y})^2}}, \tag{7.45}$$

where $\bar{x}$ and $\bar{y}$ denote the averages and the normalization by the respective standard deviations leads to $C_{xy} \in [-1, 1]$. Since C_{xy} describes a measure of linear correlation, then $C_{xy} = +1$ implies that an increase in x (with respect to its average) is accompanied by an increase in y, and vice versa. Conversely, if $C_{xy} = -1$ an increase in x takes place along with a decrease in y, and $C_{xy} = 0$ means that fluctuations of such variables are uncorrelated. The Pearson correlation coefficient has been widely used as a valuable tool to infer linear correlations between variables or sets of data in virtually all areas of science [42, 43].

We now return to the definition of the fluctuation variables $y_{\alpha\tau}(k) \equiv I_{\alpha+\tau}(k) - cI_\alpha(k)$. Here we set $c = 0$ if $\tau = 0$ and $c = 1$ if $\tau \geq 1$. Its fluctuation around the average is $\delta_{\alpha\tau}(k) = y_{\alpha\tau}(k) - \bar{y}_\tau(k)$. We thus introduce the photonic Pearson coefficient to measure the correlation between δ-fluctuations at a fixed time interval τ:

$$Q_{\alpha\beta,\tau} = \frac{\sum_k \delta_{\alpha\tau}(k)\delta_{\beta\tau}(k)}{\sqrt{\sum_k \delta^2_{\alpha\tau}(k)}\sqrt{\sum_k \delta^2_{\beta\tau}(k)}}. \tag{7.46}$$

Remarkably, this Pearson correlation coefficient is simultaneously sensitive to both turbulence and spin-glass photonic behaviors. Indeed, for $\tau = 0$ it recovers the Parisi overlap parameter of the photonic glassy phase (see Chapters 5, 6, and 8 for details). On the other hand, for $\tau \geq 1$ the Pearson coefficient is able to infer the turbulence phenomenon, as shown below. For example, the maximum turbulence-like signature (intermittency) is manifested at the shortest time scale $\tau = 1$. In contrast, for $\tau \gg 1$ the short-time-scale effects and intermittency rapidly fade away and a crossover takes place to the nonturbulent (Gaussian) behavior, as demonstrated in the following in terms of the statistical analysis of the normalized intensity increments, $x_{\alpha\tau}(k) = \delta I_{\alpha\tau}(k)/[\sum_\alpha \delta I^2_{\alpha\tau}(k)]^{1/2}$.

For the shortest time scale $\tau = 1$, Fig. 7.6(a) shows an excerpt of the large dataset, with 3000 (out of 150,000) intensity increments between successive spectra [39]. For the whole dataset, the distribution $P(x_1)$ at $\lambda = 1539.8$ nm around the peak emission, given by Eq. (7.10) with δI_1 replaced by x_1, is depicted in squares in Fig. 7.6(b). In this semi-log plot, the large deviation from the parabolic shape signalizes the non-Gaussian turbulence-like behavior of $P(x_1)$. The red solid line shows a remarkable agreement with the results of the hierarchical model. The best fit is obtained for $N = 6$ time scales of intensity fluctuations, with $P(x_1)$ given by a statistical mixture of two functions, $P_a(x_1)$ and $P_b(x_1)$: $P(x_1) = p\ P_a(x_1) + (1-p)\ P_b(x_1)$, with parameters $p = 0.30$, $\beta_a = 8.53$, $\beta_b = 6.47$, $\epsilon_{0,a} = 0.16$ and $\epsilon_{0,b} = 1.36$. Moreover, each mixture component displays non-Gaussian asymptotic behavior in the form of a heavy stretched-exponential tail, consistently with the turbulent-like state (see previous sections).

Figure 7.6(c) presents the associated variance distribution $f(\epsilon_1)$ which is also given by a statistical mixture with the same parameters as $P(x_1)$. Remarkably, as seen in Fig. 7.6(c) the effect of the small and large variances ϵ_1 are distinguishably incorporated by each mixture component. We may argue that the mixture could arise from a subtle combination of stimulated and spontaneous turbulent emissions.

On the other hand, by considering a hierarchical model similar to that described in the previous sections for the intensity increments, the distribution $P(Q_1)$ of the Pearson coefficient for $\tau = 1$ can be

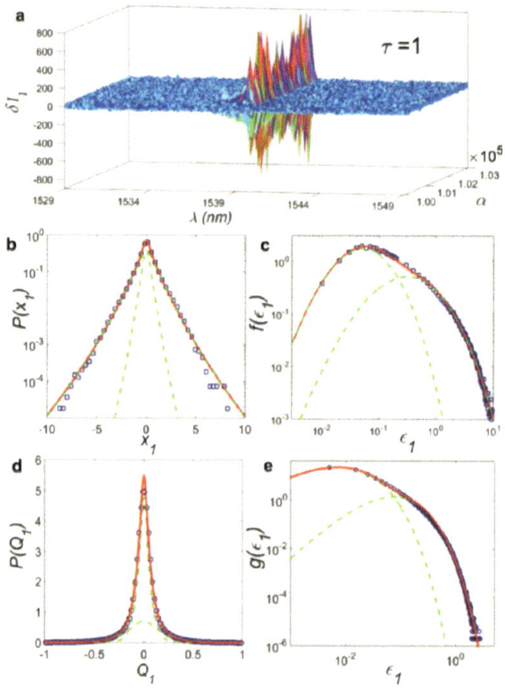

Figure 7.6 (a) Plot of 3000 experimental intensity increments, $\delta I_{\alpha\tau}(k) = I_{\alpha+\tau}(k) - I_{\alpha}(k)$, between successive spectra ($\tau = 1$) as a function of the spectrum label α and wavelength λ indexed by k in the interval [1529.0 nm, 1549.0 nm]. (b) Distribution $P(x_\tau)$, $\tau = 1$, of normalized increments $x_{\alpha\tau}(k) = \delta I_{\alpha\tau}(k)/[\sum_\alpha \delta I^2_{\alpha\tau}(k)]^{1/2}$, at the wavelength $\lambda = 1539.8$ nm around the peak emission, obtained from the whole set of 150,000 spectra (blue squares). The large deviation from the parabolic shape observed in this semi-log plot is consistent with the non-Gaussian photonic turbulent-like behavior of $P(x_1)$, which arises for this maximum intermittency state at $\tau = 1$. The red solid line displays the nice fit to the results of the hierarchical model, with $P(x_1)$ given by a statistical mixture of two Meijer-G functions with $N = 6$ intensity fluctuation time scales. (c) Associated variance distribution $f(\epsilon_1)$. Experimental data (blue squares) are also nicely fitted by the theoretical result (red solid line). The individual mixture components are shown by green dashed lines in (b) and (c). (d) Distribution $P(Q_1)$ of values of the photonic Pearson coefficient and (e) associated variance distribution $g(\epsilon_1)$ calculated from 10,001 spectra. The model results for $N_Q = 2$ (red solid lines) also adjust remarkably well to the experimental data (blue circles), with the mixture components shown in green dashed lines. (Reproduced with permission from Ref. [39].)

also calculated, yielding

$$P(Q_\tau) = \frac{c}{(1-Q_\tau^2)} G^{N^{(Q)}+1,0}_{0,N^{(Q)}+1}\left(\begin{matrix} - \\ \boldsymbol{\beta}^{(Q)} - \mathbf{1/2}, 0 \end{matrix} \middle| \frac{\omega^{(Q)} z_\tau^2}{2\epsilon_0^{(Q)}} \right), \quad (7.47)$$

with $z_\tau = 2\text{arctanh}(Q_\tau)$, $c^2 = 4\omega^{(Q)}/[2\pi\epsilon_0^{(Q)}\Gamma^2(\boldsymbol{\beta}^{(Q)})]$, and $\epsilon_0^{(Q)}$, $\omega^{(Q)}$ and $\beta^{(Q)}$ defined similarly to Eq. (7.10).

In Figs. 7.6(d–e) the experimental data (circles) are plotted together with the best theoretical fits (red lines) for $P(Q_1)$ and the associated variance distribution $g(\epsilon_1)$. A very nice agreement is observed for $N^{(Q)} = 2$ time scales of the Pearson coefficient and parameters of a similar statistical mixture: $p^{(Q)} = 0.70$, $\beta_a^{(Q)} = 2.53$, $\beta_b^{(Q)} = 3.50$, $\epsilon_{0,a}^{(Q)} = 0.03$ and $\epsilon_{0,b}^{(Q)} = 0.18$. In particular, we argue below that the unimodal behavior of $P(Q_1)$ is intrinsically related to the maximum intermittency effect underlying the turbulent-like state for $\tau = 1$.

A remarkably distinct picture emerges for large separation time scales between the emission spectra, $\tau \gg 1$. Indeed, Fig. 7.7 displays the crossover to the nonturbulent behavior for $\tau = 5000$. The intensity increments at the wavelength $\lambda = 1539.8$ nm now assume a Gaussian distribution $P(x_{5000})$, Fig. 7.7(b), indicating that intermittency has been fully suppressed (i.e., $N = 0$ in the theoretical model). Concurrently, a bimodal profile of $P(Q_{5000})$ is observed in Fig. 7.7(c), showing a significant contrast with the unimodal profile of $P(Q_1)$ that can be understood as follows. The intermittency effect tends to increase for short scales the probability of events that are rarer at much larger scales. As seen in Fig. 7.7(c), when intermittency is absent in the nonturbulent state for $\tau = 5000$, values $Q_{5000} \approx 0$ are less probable than those around $Q_{5000} = \pm 1$. Consequently, when intermittency is maximum for $\tau = 1$, values $Q_1 \approx 0$ have their probability much increased, leading to the unimodal behavior of $P(Q_1)$ in the turbulent-like state.

Finally, we also observe that the bimodal behavior of $P(Q_{5000})$ resembles the spin-glass profile [40] of the distribution $P(q)$ of the Parisi overlap parameter (see, e.g., Chapters 6 and 8) for $P/P_{\text{th}} > 1$. In fact, since the Parisi parameter considers *all* separation times τ between spectra, we thus expect that the statistical weight of the replica overlaps with $\tau \gg 1$ dominates over the long time series. As a consequence, $P(Q_\tau)$ for $\tau \gg 1$ should actually appear qualitatively

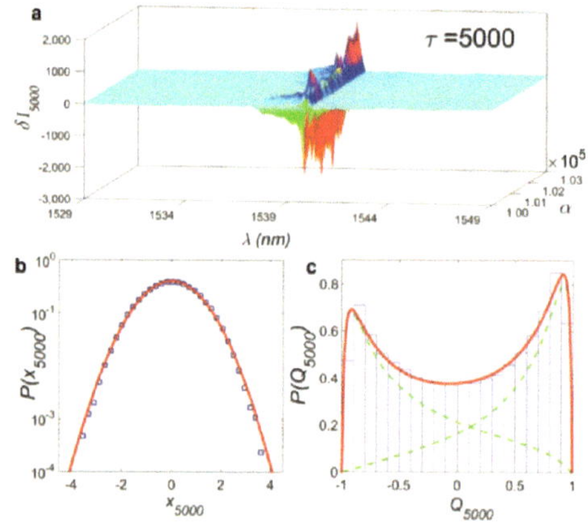

Figure 7.7 (a) Plot of 3000 experimental intensity increments, $\delta I_{\alpha\tau}(k) = I_{\alpha+\tau}(k) - I_{\alpha}(k)$, between spectra separated by $\tau = 5000$ time units, as a function of the spectrum label α and wavelength λ indexed by k in the interval [1529.0 nm, 1549.0 nm]. (b) Semi-log plot of the distribution $P(x_{\tau})$, $\tau = 5000$, of normalized increments, at the wavelength $\lambda = 1539.8$ nm around the peak emission, obtained from the whole set of 150,000 spectra (blue squares). The red solid line displays the nice fit to the results of the hierarchical model. At the large time scale $\tau = 5000$, the parabolic shape indicates that the distribution $P(x_{5000})$ assumes a Gaussian form ($N = 0$ in the theoretical model), showing that intermittency has been fully suppressed. (c) Histogram with the distribution $P(Q_{5000})$ of values of the photonic Pearson coefficient calculated from 10,001 spectra displaying a spin-glass-like bimodal profile. The experimental histogram is also nicely fitted by the theoretical model (red solid line), with the mixture components of $P(Q_{5000})$ depicted by green dashed lines. Since the intermittency rapidly fades away for $\tau > 1$, then $P(Q_{5000})$ appears qualitatively similar to the distribution $P(q)$ of the Parisi overlap parameter that characterizes the spin glass behavior. (Reproduced with permission from Ref. [39].)

similar to the distribution of the Parisi parameter that characterizes the spin glass state.

7.5 Simultaneous Evaluation of Intermittency Effects, RSB and Modes Correlations in a RL

In the previous section, we defined a Pearson correlation coefficient, Eq. (7.46), and applied it to the simultaneous analysis of the photonic intermittency and Parisi parameter of the RSB spin-glass phase in a RL system. We now introduce another Pearson coefficient to account for the simultaneous evaluation of RSB spin-glass and modes correlation behaviors [44, 45],

$$P_{MN} = \overline{\Delta_{MK}\Delta_{NK}}. \tag{7.48}$$

Above a sum over K is implicit (Einstein summation notation), $\overline{\Delta_{\gamma i}} = \Delta_{\gamma i}/\sum_K \sqrt{(\Delta_{\gamma i})^2}$, with $\Delta_{\gamma i} = I_{\gamma i} - \langle I \rangle_i$, $I_{\gamma i}$ denotes the intensity emitted by the RL at the wavelength λ_i in the spectrum γ, and $K = i$ ($K = \gamma$) representing the wavelength λ_i (spectrum γ). The advantage to use this modified Pearson coefficient is that it comprises both the Parisi overlap parameter, Eq. (7.46), if $M = \alpha$ and $N = \beta$, as well as the standard Pearson coefficient to measure correlation between intensity fluctuations at wavelengths λ_i and λ_j if $M = i$ and $M = j$,

$$C_{ij} = \frac{\sum_\gamma \Delta_{\gamma i}\Delta_{\gamma j}}{\sqrt{\sum_\gamma (\Delta_{\gamma i})^2}\sqrt{\sum_\gamma (\Delta_{\gamma j})^2}}. \tag{7.49}$$

Thus, by employing Eqs. (7.46) and (7.48) it is possible to analyze jointly the intertwined effects of turbulence-like intermittency, RSB spin-glass, and modes correlation behaviors in a given RL system.

The first use of the Pearson coefficient to evaluate correlations between modes in RLs was in the study by Leonetti and coauthors [46] of the emission properties of a nanofiber RL system with and without TiO_2 nanoparticle scatterers, whereas in the work by Montinaro and coauthors [47] the mode-locking transition was investigated by varying the excitation beam profile in a RL with TiO_2 nanoparticles in Rhodamine dye solution. More recently, Sciuti and coworkers [48] studied the modes dynamics in the mode-locking transition via the temporal mapping of the emitted spectra characterized by the standard Pearson correlation in a RL based on dye-doped electrospun nanofibers. Also, Sarkar and coworkers related [49] the modes correlations to the RSB and Lévy-like properties of a weakly scattering optofluidic RL employing a covariance measure with a slightly distinct normalization. Pearson coefficients have been also recently applied by our group in the above modified versions to other RL systems [44, 45], including a hybrid electronically addressable random (HEAR) laser [45], which is a novel type of RFL with the remarkable property of selection of the fiber section with lasing emission [50].

Below we shall focus on the intermittency effects, RSB and modes correlations in a Nd:YAG RL synthesized using a spray pyrolysis method (see [44] for experimental details). As for the joint analysis of RSB spin-glass and intermittency properties in this RL, quite similar results to those discussed in Figs. 7.6 and 7.7 have been found [44]. So, below we concentrate on the analysis of the RSB spin-glass and modes correlation behaviors using Eq. (7.49).

Figure 7.8 shows [44] three sets of vertical panels. The ones on the left column show the distribution $P(q)$ of values of the Parisi overlap parameter; in the center we depict the profiles of 1000 recorded spectra; and on the right column we present results for the Pearson correlation coefficient C_{ij}, Eq. (7.49). In particular, in the latter all cross-correlations between intensity fluctuations at distinct wavelengths, as well as self-correlations (thin diagonal red lines), are shown in color scale in the form of heatmap plots.

Each horizontal line in Fig. 7.8 corresponds to a value of the excitation power, from the prelasing regime well below threshold, with $P/P_{\rm th} = 0.18$ and $P/P_{\rm th} = 0.56$ (two upper lines), to the RL phase around the threshold with $P/P_{\rm th} = 1.32$ (center line), and the RL regime well above threshold, $P/P_{\rm th} = 2.07$ and $P/P_{\rm th} = 2.45$ (two lower lines).

We first note in Figs. 7.8(b) and 7.8(e) that the spectra below threshold present rather weak intensity fluctuations in association with mostly blue heatmap patterns in the Pearson coefficient (Figs. 7.8(c) and 7.8(f)), evidencing that intensity fluctuations are fairly uncorrelated in the prelasing state. Consistently, the unimodal distributions $P(q)$ around $q = 0$, seen in Figs. 7.8(a) and 7.8(d), indicate a replica-symmetric regime.

As the excitation power is increased and the threshold is approached, we observe in the central line of Fig. 7.8 the concomitant onset of correlations (orange and red in the heatmap of the Pearson coefficient C_{ij}, Fig. 7.8(i)), along with larger fluctuations, Fig. 7.8(h). A simultaneous emergence of the photonic spin-glass RSB phenomenon takes place, as indicated in Fig. 7.8(g) by the presence of the side peaks in the bimodal $P(q)$.

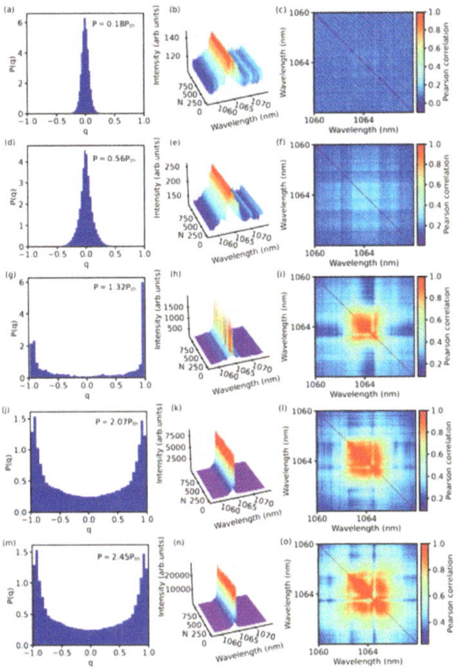

Figure 7.8 RSB behavior and modes correlations in the Nd:YAG RL. Distribution $P(q)$ of Parisi overlap parameter values (left column), Eq. (7.46), 1000 spectra profiles (center), and heatmap plots of the Pearson correlation coefficient (right column), Eq. (7.49). Each horizontal line corresponds to an excitation power, from below threshold (two upper lines: $P/P_{\mathrm{th}} = 0.18$ and $P/P_{\mathrm{th}} = 0.56$), to around (central line: $P/P_{\mathrm{th}} = 1.32$), and well above threshold (two lower lines: $P/P_{\mathrm{th}} = 2.07$ and $P/P_{\mathrm{th}} = 2.45$). Below threshold, a replica-symmetric paramagnetic-like phase identified by a unimodal central peaked $P(q)$ coincides with weakly fluctuating spectra and uncorrelated modes with mostly blue heatmaps of the Pearson coefficient. Above threshold, the RSB glassy behavior sets in with two side peaks in $P(q)$ along with stronger intensity fluctuations and increasingly correlated modes (orange and red regions in the heatmaps). (Reproduced with permission from Ref. [44].)

Finally, as the excitation power is enhanced further (last two lines of Fig. 7.8), fluctuations nail down (Figs. 7.8(k) and 7.8(n)) in the RSB regime above threshold displaying a bimodal $P(q)$ (Figs. 7.8(j) and 7.8(m)), with even stronger correlations between modes (larger red regions in Figs. 7.8(l) and 7.8(o)).

7.6 Concluding Remarks

To conclude this chapter, we comment that the joint analysis of the photonic turbulence-like properties, spin-glass phase with RSB, and dynamics correlations between lasing modes through a couple of modified Pearson correlation coefficients shows how intertwined these behaviors are in RL systems, and suggest that they might share common underlying mechanisms.

We thus hope the techniques presented in this chapter can stimulate the quest for the joint understanding of these complex behaviors in RL systems under a unified theoretical approach.

Acknowledgments

We thank the financial support from the Brazilian agencies Conselho Nacional de Desenvolvimento Científico e Tecnológico (CNPq), Coordenação de Aperfeiçoamento de Pessoal de Nível Superior (CAPES), and Fundação de Amparo à Ciência e Tecnologia do Estado de Pernambuco (FACEPE).

References

1. Callies, J., Ferrari, L., Klymak, J. M., and Gula, J. Seasonality in submesoscale turbulence. *Nature Commun.* **6**, 6862 (2015).
2. Sasaki, H., Klein, P., and Sasai, Y. Impact of oceanic-scale interactions on the seasonal modulation of ocean dynamics by the atmosphere. *Nature Commun.* **5**, 5636 (2014).
3. Grošelj, D., Jenko, F., and Frey, E. How turbulence regulates biodiversity in systems with cyclic competition. *Phys. Rev. E* **91**, 033009 (2015).
4. Schmitt, J. M., and Kumar, G. Turbulent nature of refractive-index variations in biological tissue. *Opt. Lett.* **21**, 1310–1312 (1996).
5. Turitsyna, E. G., Smirnov, S. V., Sugavanam, S., Tarasov, N., Shu, X., Babin, S. A., Podivilov, E. V., Churkin, D. V., Falkovich, G., and Turitsyn, S. K. The laminar turbulent transition in a fibre laser. *Nature Photon.* **7**, 783–786 (2013).

6. Wabnitz, S. Optical turbulence in fiber lasers. *Opt. Lett.* **39**, 1362–1364 (2014).
7. Henn, E. A. L., Seman, J. A., Roati, G., Magalhães, K. M. F., and Bagnato, V. S. Emergence of turbulence in an oscillating Bose–Einstein condensate. *Phys. Rev. Lett.* **103**, 045301 (2009).
8. Xu, G., Vocke, D., Faccio, D., Garnier, J., Roger, T., Trillo, S., and Picozzi, A. From coherent shocklets to giant collective incoherent shock waves in nonlocal turbulent flows. *Nature Commun.* **6**, 8131 (2015).
9. Laurie, J., Bortolozzo, U., Nazarenko, S., and Residori, S. One-dimensional optical wave turbulence: experiment and theory. *Phys. Rep.* **514**, 121–175 (2012).
10. Korotkova, O. *Random Light Beams: Theory and Applications* (CRC Press, Boca Raton, 2014).
11. Ghashghaie, S., Breymann, W., Peinke, J., Talkner, P., and Dodge, Y. Turbulent cascades in foreign exchange markets, *Nature* **381**, 767–770 (1996).
12. Lux, T. Turbulence in financial markets: the surprising explanatory power of simple cascade models. *Quant. Financ.* **1**, 632–640 (2010).
13. Malik, M., O'Sullivan, M., Rodenburg, B., Mirhosseini, M., Leach, J., Lavery, M. P., Padgett, M. J., and Boyd, R. W. Influence of atmospheric turbulence on optical communications using orbital angular momentum for encoding. *Opt. Express* **20**, 13195–13200 (2012).
14. Frisch, U. *Turbulence: The Legacy of A. N. Kolmogorov* (Cambridge University Press, Cambridge, 1995).
15. Siegman, A. E. *Lasers* (University Science Books, Mill Valley, 1986).
16. See *Focus Issue* on *Fibre Lasers, Nature Photon.* **7**, 841–932 (2013).
17. Luan, F., Gu, B., Gomes, A. S. L., Yong, K.-T., Wen, S., and Prasad, P. N. Lasing in nanocomposite random media. *NanoToday* **10**, 168–192 (2015).
18. Wiersma, D. S. Disordered photonics. *Nature Photon.* **7**, 188–196 (2013).
19. Sosa-Correa, W., Pereira, R. M., Macêdo, A. M. S., Raposo, E. P., Salazar, D. S. P., & Vasconcelos, G. L., Emergence of skewed non-Gaussian distributions of velocity increments in isotropic turbulence, *Phys. Rev. Fluids* **4** 064602 (2019).
20. González, I. R. R., Lima, B. C., Pincheira, P. I. R., Brum, A. A., Macêdo, A. M. S., Vasconcelos, G. L., Menezes, L. de S., Raposo, E. P., Gomes, A. S. L., and Kashyap, R. Turbulence hierarchy in a random fibre laser. *Nat. Commun.* **8**, 15731 (2017).

21. Beck C. Dynamical foundations of nonextensive statistical mechanics, *Phys. Rev. Lett.* **87**, 180601 (2001).
22. Salazar, D. S. P., and Vasconcelos, G. L. Stochastic dynamical model of intermittency in fully developed turbulence. *Phys. Rev. E* **82**, 047301 (2010).
23. Macêdo, A. M. S., González, I. R. R., Salazar, D. S. P., and Vasconcelos, G. L. Universality classes of fluctuation dynamics in hierarchical complex systems. *Phys. Rev. E* **95**, 032315 (2017).
24. Mathai A. M., Saxena R. K., & Haubold H. J. *The H-Function: Theory and Applications* (Springer-Verlag, New York, 2009).
25. Vasconcelos, G. L., Salazar, D. S. P., and Macêdo, A. M. S. Macêdo, A. M. S. Maximum entropy approach to H-theory: Statistical mechanics of hierarchical systems. *Phys. Rev. E* **97**, 022104 (2018).
26. Van den Broeck C.& Esposito M. Three faces of the second law. II. Fokker–Planck formulation *Phys. Rev. E* **82**, 011144 (2010).
27. Tomé T., and de Oliveira M. J. Entropy production in irreversible systems described by a Fokker–Planck equation. *Phys. Rev. E* **82**, 021120 (2010).
28. San Liang X. entropy evolution and uncertainty estimation with dynamical systems. *Entropy* **16**, 3605 (2014).
29. Abe S. Fokker–Planck theory of nonequilibrium systems governed by hierarchical dynamics *Found. Phys.* **44**, 175 (2014).
30. Macêdo A. M. S., González I. R. R., Raposo E. P., Menezes L. de S., and Gomes A. S. L., Turbulent intermittency in a random fiber laser *Atoms* **7(2)**, 43 (2019).
31. Jakeman, E., and Pusey, P. N. Significance of *K* distributions in scattering experiments. *Phys. Rev. Lett.* **40**, 546–550 (1978).
32. Van Kampen, N.G. *The master equation. Stochastic Processes in Physics and Chemistry, 3rd ed.* (Elsevier: Amsterdam, The Netherlands, 2007).
33. Redding, B., Choma, M. A., and Cao, H. Speckle-free laser imaging using random laser illumination. *Nature Photon.* **6**, 355–359 (2012).
34. Cao, H. Random thoughts. *Nature Photon.* **7**, 164 (2013).
35. Lepri S., Cavalieri S., Oppo G. L., & Wiersma D. S. Statistical regimes of random laser fluctuations. *Phys. Rev. A* **75**, 063820 (2007).
36. Falkovich, G. Introduction to turbulence theory, in *Non-Equilibrium Statistical Mechanics and Turbulence* (Cambridge University Press, Cambridge, 2008). Edited by Nazarenko, S., and Zaboronski, O. V.

37. Ward, K., Tough, R., and Watts, S. *Sea Clutter: Scattering, the K Distribution and Radar Performance* (The Institution of Engineering and Technology, Herts, 2013).
38. Yamazaki, H., Lueck, R. G., and Osborn, T. A comparison of turbulence data from a submarine and a vertical profiler. *J. Phys. Oceanogr.* **20**, 1778–1786 (1990).
39. González I. R. R., Raposo E. P., Macêdo A. M. S., Menezes L. de S., & Gomes A. S. L. Coexistence of turbulence-like and glassy behaviours in a photonic system, Sci. Rep. **8**, 17046 (2018).
40. Gomes, A. S. L., Lima, B. C., Pincheira, P. I. R., Moura, A. L., Gagné, M., Raposo, E. P., de Araújo, C. B., and Kashyap, R. Glassy behavior in a one-dimensional continuous-wave erbium-doped random fiber laser. *Phys. Rev. A* **94**, 011801(R) (2016).
41. Lima, B. C., Gomes, A. S. L., Pincheira, P. I. R., Moura, A. L., Gagné, M., Raposo, E. P., de Araújo, C. B., and Kashyap, R. Observation of Lévy statistics in one-dimensional erbium-based random fiber laser, *J. Opt. Soc. Am. B* **34**, 293–299 (2017).
42. Pearson, k. Note on regression and inheritance in the case of two parents, *Proc. Roy. Soc. Lond.* **58**, 240–242 (1895).
43. Rodgers, J. L., and Nicewander, W. A. Thirteen ways to look at the correlation coefficient, *Am. Statistician* **42**, 59–66 (1988).
44. Coronel, E., da Silva-Neto, M. L., Moura, A. L., González, I. R. R., Pugina, R. S., Hilário, E. G., da Rocha, E. G., Caiut, J. M. A., Gomes, A. S. L., and Raposo, E. P. Simultaneous evaluation of intermittency effects, replica yymmetry breaking and modes dynamics correlations in Nd:YAG random lasers, submitted (2021).
45. Coronel, E., Das, A., González, I. R. R., Gomes, A. S. L., Margulis, W., von der Weid, J. P., and Raposo, E. P. Evaluation of Pearson correlation coefficient and Parisi parameter of replica symmetry breaking in a hybrid electronic addressable random fiber laser, *Opt. Express* **29**, 24422–24433 (2021).
46. Leonetti, M., Conti, C., and Lopez, C. The mode-locking transition of random lasers, *Nat. Photon.* **5**, 615–617 (2011).
47. Montinaro, M., Resta, V., Camposeo, A., Moffa, M., Morello, G., Persano, L., Kazlauskas, K., Jursenas, S., Tomkeviciene, A., Grazulevicius, J. V., and Pisignano, D. Diverse regimes of mode intensity correlation in nanofiber random lasers through nanoparticle doping, *ACS Photon.* **5**, 1026–1033 (2018).

48. Sciuti, L. F., Mercante, L. A., Correa, D. S., and De Boni L. Random laser in dye-doped electrospun nanofibers: Study of laser mode dynamics via temporal mapping of emission spectra using Pearson's correlation, *J. Lumin.* **224**, 117281 (2020).
49. Sarkar, A., Andreasen, J., and Bhaktha, B. B. N. S. Replica symmetry breaking in a weakly scattering optofluidic random laser, *Sci. Rep.* **10**, 2628 (2020).
50. Margulis, W., Das, A., von der Weid, J. P., and Gomes, A. S. L. Hybrid electronically addressable random fiber laser, *Opt. Express* **28**, 23388 (2020).

Chapter 8

Floquet Spin-Glass Phase in a Random Fiber Laser

Ernesto P. Raposo,[a] Iván R. R. González,[a] A. M. S. Macêdo,[a] Leonardo de S. Menezes,[b] and Anderson S. L. Gomes[b]
[a]*Laboratório de Física Teórica e Computacional, Departamento de Física, Universidade Federal de Pernambuco, 50670-901 Recife, Pernambuco, Brazil*
[b]*Departamento de Física, Universidade Federal de Pernambuco, 50670-901 Recife, Pernambuco, Brazil*
ernesto@df.ufpe.br

8.1 Introduction

The pace of relevant discoveries in random laser (RL) and random fiber laser (RFL) systems has not slowed down in recent years. On the contrary, it seems that novel interesting phenomena are on the threshold of being unveiled in such systems, both from the experimental and theoretical points of view.

In this chapter, we discuss the example of a remarkable new photonic phase that has been very recently reported in a strongly disordered erbium-based RFL, the so-called Floquet spin-glass

Lévy Statistics and Spin Glass Behavior in Random Lasers
Edited by Anderson S. L. Gomes, André L. Moura, Cid B. de Araújo, and Ernesto P. Raposo

ISBN 978-981-4968-55-3 (Hardcover), 978-1-003-33618-1 (eBook)
www.jennystanford.com

phase [1]. Differently from the photonic *equilibrium* spin-glass phase reported in several RLs and RFLs (see, e.g., Chapters 5–7 of this book), the Floquet glassy behavior in the erbium-based RFL is clearly a *nonequilibrium* one. As described in the following, its mechanistic origin can be traced back to the periodic oscillations in time of the quadratic and quartic couplings in the system's photonic Hamiltonian, which, in turn, are related to a periodic oscillation of the continuous-wave (CW) source power. Floquet phases have been intensely investigated in the past few years, with considerable theoretical advances but relatively few evidences reported in actual experimental systems. The Floquet phase discussed in the following is the first one to be discovered in a system which employs a laser emission-based mechanism.

It is well known that thermalization and phase structure of interacting systems are central tenets of equilibrium (either classical or quantum) physics (see, e.g., Ref. [2] and references therein). In spite of this, the fundamental quest for the extension of the phase structure concept to out-of-equilibrium systems is still an open issue. Such debate has ignited a recent burst of interest [2–28] in quantum disordered many-body interacting Floquet systems described by a periodically driven Hamiltonian. Indeed, the rather nontrivial nonequilibrium Floquet phases conceivably represent the maximum possible dynamical extension of the canonical notion of phase structure in general physical systems [2].

Nonequilibrium many-body interacting systems in contact with some sort of driving source are generally expected to absorb energy indefinitely, thus heating up towards a trivial "infinite temperature" state [7]. In this sense, two essential conditions make driven quantum Floquet systems an entirely special case: disorder and localization. A suitable amount of disorder can lead to spatially localized eigenmodes that couple weakly with faraway modes. In such a situation, the energy transfer between the system modes is hampered and the absorption from the driving source saturates, therefore preventing the route to the "infinite temperature" state, a phenomenon known as many-body localization. It is then acknowledged that disorder can eventually take a remarkable stabilizing role for the dynamical Floquet phases.

On the theoretical side, a number of Floquet systems with nontrivial phase structure have been recently reported, for instance, in quantum disordered spin Hamiltonians with periodic couplings between spins, which display a magnetic spin-glass phase [2–11]. However, the experimental counterpart scenario seems to advance in a relatively slower pace. Indeed, in recent years, very few systems have been demonstrated to present experimental Floquet phases, most of them regarding the so-called discrete time crystals [26–28] and none employing a laser emission-based mechanism, such as in the RFL one described below [1]. In fact, some experimental realizations of Floquet phases involve the observation of discrete time crystals in periodically driven disordered spin chains of trapped ions [27], as well as in an extensive disordered ensemble of dipolar spin impurities in diamond at room temperature [28].

Below we review the first experimental evidence of a Floquet spin-glass phase observed in a photonic system [1], a quasi-one-dimensional erbium-based RFL with strong embedded structural disorder. Disorder, eigenmode localization, and nonlinearity play essential roles to stabilize this photonic Floquet phase, as mentioned.

RLs and RFLs, described as open complex photonic systems, have been recently employed as successful platforms to access nontrivial complex phases, such as equilibrium spin glasses [29–36] and turbulent regimes [37–40], as theoretically justified in [38–42] (for more details, we refer the reader to Chapters 5–7 of this book). In the erbium-doped RFL studied in the present work, disorder was introduced by inscribing a remarkable amount ($\sim 10^3$) of specially designed stationary (i.e., quenched) random Bragg gratings [43] in the fiber core. Different from conventional lasers with fixed mirrors employed to sustain the laser oscillations, the disordered Bragg gratings in this RFL act as random scatterers of photons in the laser medium with gain provided by the erbium ions, thus generating the required feedback for the random laser emission above the RFL threshold. At this point, it is important to remark that by using the speckle contrast technique, the erbium-based RFL was shown [34] to emit a large number (>200) of longitudinal modes, which confers its quasi-one-dimensional character [43]. The quenched disorder is also responsible for the spatial localization of the RFL modes [43] that interact weakly with distant modes (see below). These combined ingredients prevent the coherent oscillation of modes, so that a quasi-equilibrium photonic spin glass phase was reported above the RFL threshold [34–36], with replica symmetry breaking (RSB) properties characterized under Parisi's framework to study magnetic spin glasses [44, 45].

In the next section, we present the theoretical background underlying the description of the erbium-based RFL in terms of a quantum system-and-bath Floquet Hamiltonian, with periodic CW pump intensity driving the random quadratic and quartic couplings between the lasing modes. In Section 8.3, the experimental analysis is detailed, with the RFL submitted to a large number of periodic oscillations of the drive and the correlation statistics of a rather extensive number (150,000) of RFL emission spectra well above threshold is analyzed. In the strongly nonlinear disordered regime a photonic Floquet spin-glass phase is clearly evidenced from the dynamics of the Parisi overlap parameter and asymmetry properties of the spin-glass order parameter distribution function. In contrast, no Floquet phase can be characterized in the prelasing regime below threshold when mode localization is absent. Finally, concluding remarks are left to Section 8.4.

8.2 Theoretical Background

We begin by reviewing the derivation of the quantum system-bath Hamiltonian of the erbium-based RFL in the form

$$\mathcal{H} = \mathcal{H}_{\mathrm{S}} + \mathcal{H}_{\mathrm{B}} + \mathcal{H}_{\mathrm{SB}}, \tag{8.1}$$

where the first two terms above describe, respectively, the quantized electromagnetic field inside and outside the RFL system, while the term $\mathcal{H}_{\mathrm{SB}}$ accounts for the system-bath interaction (see, also, Chapter 5). Next, the connection with the Floquet formalism will be built by introducing periodicity in time so that $\mathcal{H}_{\mathrm{S}}(t) = \mathcal{H}_{\mathrm{S}}(t+T)$, as explained below.

By following closely the approach in [46, 47], we use the Feshbach projection technique to separate the contributions to the electromagnetic field from the regions inside and outside the RFL, yielding (in $\hbar$ units)

$$\mathcal{H}_{\mathrm{S}} = \sum_{k} \omega_k a_k^{\dagger}(t) a_k(t),$$

$$\mathcal{H}_{\mathrm{B}} = \int d\omega \sum_{m} \omega b_m^{\dagger}(\omega, t) b_m(\omega, t),$$

$$\mathcal{H}_{\mathrm{SB}} = \int d\omega \sum_{k,m} \mathcal{W}_{km}(\omega) a_k^{\dagger}(t) b_m(\omega, t) + \text{h.c.}, \tag{8.2}$$

where $a_k^{\dagger}(t)$ and $a_k(t)$ are creation and annihilation operators obeying canonical commutation relations in the Heisenberg representation, which are assigned to the field eigenmodes of the RFL with discrete eigenfrequencies ω_k indexed by k. Analogously, $b_m^{\dagger}(\omega, t)$ and $b_m(\omega, t)$ form a continuous set of operators for the modes outside the RFL with frequency ω and channel discrete index m related, e.g., to the expansion of the external solutions in angular momentum states [46]. Thus, $\mathcal{H}_{\mathrm{B}}$ describes a large reservoir of quantum oscillators interacting bilinearly with the RFL operators, as indicated in $\mathcal{H}_{\mathrm{SB}}$ by the coupling amplitudes $\mathcal{W}_{km}(\omega)$ in the rotating-wave approximation [46].

The Heisenberg equations of motion that follow from Eq. (8.2) read

$$\frac{da_k}{dt} = -i\omega_k a_k - i \int d\omega \sum_{m} \mathcal{W}_{km} b_m,$$

$$\frac{db_m}{dt} = -i\omega b_m - i \sum_{k} \hbar \mathcal{W}_{km}^{*} a_k. \tag{8.3}$$

Next, in order to account for the nonlinear effects arising from the gain provided by the erbium ions in the RFL system, the Jaynes–Cummings atom-field interaction [48, 49] has to be added to $\mathcal{H}_{\mathrm{S}}$. This implies the introduction of the term $\int d\mathbf{r} g_k^{*}(\mathbf{r}) \sigma_{-}(\mathbf{r}, t)$ in the right hand side of the da_k/dt dynamics in Eq. (8.3), with the integral performed over the RFL region and σ_{-} denoting the atomic density lowering operator for the erbium ions. The atom-field coupling $g_k(\mathbf{r})$ is proportional to $\mathbf{p} \cdot \mathbf{E}_k(\mathbf{r})$, where $\mathbf{p}$ represents the atomic dipole matrix elements and $\mathbf{E}_k(\mathbf{r})$ is the electric field amplitude inside the RFL. We also denote by $n(\mathbf{r})$ the RFL refractive index which carries information on the quenched spatial disorder due to the random fiber Bragg gratings.

The dynamics of the operators a_k and b_m has to be considered along with the dynamics equations of σ_- and the density inversion operator $\sigma_z(\mathbf{r}, t)$, namely [47]

$$\frac{d\sigma_-}{dt} = -(\gamma_\perp + i\omega_a)\sigma_- + 2\sum_k g_k \sigma_z a_k + F_-,$$
$$\frac{d\sigma_z}{dt} = \gamma_\parallel (S\rho - \sigma_z) - \sum_k (g_k^* a_k^\dagger \sigma_- + \text{h.c.}) + F_z, \tag{8.4}$$

where $\gamma_\perp$ and $\gamma_\parallel$ denote decay rates, ω_a is the frequency of the active transition in the erbium ions, $\rho(\mathbf{r})$ is the density of erbium ions, the noise terms $F_-(\mathbf{r}, t)$ and $F_z(\mathbf{r}, t)$ arise from the coupling of the erbium ions with the external bath, and S is the pump intensity.

Since the dynamics of σ_- and σ_z evolve much faster than that of a_k and b_m, Eq. (8.4) is solved recursively in the stationary regime, and the solution for σ_- up to third order in a_k is substituted into the equation of motion for da_k/dt. Further, by considering the coupling amplitudes $\mathcal{W}_{km}$ independent of ω for a sufficiently large band around ω_a (Markov approximation [46, 50]), the final form of da_k/dt allows to write [41, 47] the effective RFL quantum Hamiltonian as

$$\mathcal{H}_\text{S} = -\frac{1}{2}\sum_{k_1 k_2} g^{(2)}_{k_1 k_2} a^\dagger_{k_1} a_{k_2}$$
$$-\frac{1}{4!}\sum_{k_1 k_2 k_3 k_4} g^{(4)}_{k_1 k_2 k_3 k_4} a^\dagger_{k_1} a_{k_2} a^\dagger_{k_3} a_{k_4}, \tag{8.5}$$

with

$$g^{(2)}_{k_1 k_2} = -2\omega_{k_1}\delta_{k_1 k_2} + 2\pi i(\mathcal{W}\mathcal{W}^\dagger)_{k_1 k_2} + G^{(2)}_{k_1 k_2} \tag{8.6}$$

and the S-dependent disordered couplings,

$$G^{(2)}_{k_1 k_2} \propto S \int d\mathbf{r}\rho(\mathbf{r}) g^*_{k_1}(\mathbf{r}) g_{k_2}(\mathbf{r}),$$
$$g^{(4)}_{k_1 k_2 k_3 k_4} \propto S \int d\mathbf{r}\rho(\mathbf{r}) g^*_{k_1}(\mathbf{r}) g_{k_2}(\mathbf{r}) g^*_{k_3}(\mathbf{r}) g_{k_4}(\mathbf{r}). \tag{8.7}$$

Likewise, one can also express

$$\mathcal{H}_\text{SB} = \sum_k i\xi_k a_k^\dagger + \text{h.c.}, \tag{8.8}$$

with

$$\xi_k(t) \equiv -i \int d\omega \sum_m e^{-i\omega(t-t_0)} \mathcal{W}_{km} b_m(\omega, t_0) \tag{8.9}$$

interpreted as a quantum noise operator [51] that depends on the bath operators only at the given time $t = t_0$ [46, 47]. Such noise can be made white and uncorrelated in a suitable basis of slow-amplitude modes [41].

The quartic coupling $g^{(4)}$ above is the main source of nonlinearity in the RFL system, since it is related to the nonlinear susceptibility $\chi^{(3)}$ arising from the process of single-photon induced nonlinear absorption, whose microscopic origin stems from the electronic levels of the erbium ions. We further remark on the disordered character of the couplings $g^{(p)}$ in $\mathcal{H}_S$, both quadratic (order $p = 2$) and quartic ($p = 4$), due to the random positions of the erbium ions and random spatial profile of the RFL refractive index.

Noticeably, a quantum system-and-bath Hamiltonian similar to that of Eq. (8.1), with a p-th order interacting $\mathcal{H}_S$, was considered in Ref. [52], in which the external bath of quantum oscillators also couples bilinearly with the system's spin operators. In this case, the Hamiltonian $\mathcal{H}_S$ defines a quantum p-spin spherical model, also possibly expressed in terms of creation and annihilation operators, with the spherical constraint on the sum of squared spin operators [52] (we notice in passing that, in the photonic context, the analogous constraint corresponds to fix the total optical power [41]). Remarkably, when random p-th order couplings with Gaussian distribution are taken into account in the Hamiltonian $\mathcal{H}_S$ with coupling orders $p > 2$, an equilibrium quantum spin-glass phase emerges [52] under Parisi's RSB formalism [44, 45].

Moreover, in the semiclassical context a Hamiltonian similar to that of Eq. (8.5), but with the operators $a_k^\dagger$ and a_k replaced by their expectation values, also yields a photonic RSB spin glass phase in multimode random laser systems in closed and open cavities, for sufficiently strong disorder in the $p = 2$ and $p = 4$ Gaussian couplings [41].

On the experimental side, the *equilibrium* photonic spin glass phase with RSB properties was first demonstrated in an amorphous solid-state RL [29] (see, also, Refs. [30–36] for more photonic experimental examples of such phase). Actually, the work by Ghofraniha and collaborators [29] constituted the first direct empirical evidence of the RSB phenomenon in any actual physical system (see, also, Chapter 6). In principle, the spin-glass order parameter should be inferred from the Parisi replica overlap parameter defined as [44, 45]

$$Q_{\alpha\beta} \propto \frac{1}{N} \sum_{k} a_{k,\alpha} a^{\dagger}_{k,\beta}, \tag{8.10}$$

with N representing the number of modes and the replica indexes α and β labeling photonic mode configurations. In the case of RL and RFL systems, such replica configurations are associated with output spectra emitted at different times under CW pumping or after distinct shots in a pulsed source. Nevertheless, since the experimental access to the phases of the operators expectation values is actually unfeasible in RL systems [42], in practice overlaps between intensity fluctuations determined from $a^{\dagger}_{k,\alpha} a_{k,\alpha}$ and $a^{\dagger}_{k,\beta} a_{k,\beta}$ are experimentally measured through the correlation parameter [29]

$$q_{\alpha\beta} = \frac{\sum_k \Delta_\alpha(k)\Delta_\beta(k)}{\sqrt{\sum_k \Delta^2_\alpha(k)}\sqrt{\sum_k \Delta^2_\beta(k)}}, \tag{8.11}$$

with intensity fluctuations $\Delta_\alpha(k) = I_\alpha(k) - \bar{I}(k)$ and $\bar{I}(k) = \sum_\alpha I_\alpha(k)/N_s$ representing the average intensity over N_s system replicas. Indeed, it was actually shown in [42] that $q_{\alpha\beta} \propto Q^2_{\alpha\beta}$ in the photonic spin-glass phase.

In the equilibrium photonic RSB spin-glass state above the RL threshold, the coherent oscillation of the nontrivially correlated modes is frustrated and the static distribution $P(q)$ of $q_{\alpha\beta}$-values displays a bimodal profile, with two essentially symmetric maxima around $q = \pm 1$ [29, 42]. In contrast, in the replica-symmetric prelasing phase below threshold modes are uncorrelated and generally only a single maximum occurs at $q = 0$, with the distribution $P(q)$ presenting negligible values at $q = \pm 1$. Therefore, the photonic RSB spin-glass transition is established when the parameter $q_{\max}$, corresponding to the position of the maximum of $P(|q|)$, shifts from zero to ≈ 1 as the pump intensity increases, a value that generally saturates for excitation powers P well above the threshold power P_{th}. In addition to the first experimental demonstration of photonic RSB spin-glass phase in a RL system, more recently other experimental evidences of equilibrium photonic

RSB spin glass phase have been also reported in several disordered materials [30–33], including the erbium-based RFL [34–36].

Having discussed the emergence of the *equilibrium* spin glass phase of the quantum system-and-bath Hamiltonian, Eq. (8.1), where $\mathcal{H}_S$ is given in Eq. (8.5), we now turn to assess its *nonequilibrium* properties particularly in the Floquet regime, in which the system's periodic properties in time are induced by some driving mechanism so that $\mathcal{H}_S(t) = \mathcal{H}_S(t + T)$.

As indicated in Eq. (8.7), one possible experimental route to achieve this is by subjecting the RFL system to a CW pump intensity $S(t)$ with a periodic slowly oscillating envelope of period T larger than the characteristic system times. It is also important to remark that the quenched disorder due to the random Bragg gratings leads to spatially localized modes in the erbium-based RFL system [43]. This mechanism of energy localization is actually essential for stabilizing the Floquet phases in driven many-body quantum systems [7]. Specifically, in the present RFL, with $\sim 10^3$ random gratings, the average transmission spectrum, given by

$$\mathrm{Tr} \sim e^{-L/\xi}, \tag{8.12}$$

implies a localization length ξ that is $\approx 6\times$ shorter than the fiber length $L = 30$ cm [43]. In this context, we also comment that the long-time Floquet states become sensitive to the system's initial conditions, in connection with the dynamical properties of the drive.

As mentioned in the Introduction section, Floquet spin-glass phases have been recently reported through theoretical studies in quantum disordered spin systems [2–11]. Remarkably, in these cases the periodic drive can induce two of such phases: an analog of the equilibrium spin glass phase, with the dynamics order parameter displaying the *same* period T of the drive, and a highly nontrivial π-spin glass with period doubling. Their characterization was made, for example, by studying the temporal oscillations of the time-dependent spin-spin correlation function, which is related to the dynamics extension of the spin-glass equilibrium order parameter [44, 45]. In the present photonic CW-pumped RFL system, the time variable t can be assigned to the mode configuration label α through $t = \alpha\tau_0$, where $\tau_0 = 100$ ms is the integration time interval between two consecutive spectra acquisitions. It is thus possible to define the dynamics extension of the Parisi overlap parameters through $Q_\alpha(\tau) \equiv Q_{\alpha,\alpha+\tau}$ and $q_\alpha(\tau) \equiv q_{\alpha,\alpha+\tau}$, in order to infer correlations between mode configurations separated in time by a fixed $\tau \equiv \Delta t = \beta - \alpha$ (in τ_0 units), and calculate the associated distributions.

In the undriven equilibrium spin-glass regime discussed above, the overlap distribution $P(q)$ is inherently time-independent, with static and symmetric maxima at $q = \pm 1$. In other words, it is the same for any statistically significant subset of configurations chosen from the whole set $\alpha, \beta = 1, \ldots, N_s$. Moreover, its maxima at $q = \pm 1$ arc static and symmetric. In contrast, in the periodically driven photonic Floquet spin glass phase the distribution $P(q)$ acquires a strong τ-dependence characterized by the oscillatory behavior of $q_{\max}$ and periodic variations of the asymmetry between the maxima at $q = -1$ and $q = +1$, as we shall see below.

8.3 Experimental Results in the Erbium-Doped RFL

We now turn to analyze the experimental results on the erbium-doped RFL system [1]. The details of the experimental setup and fiber characteristics can be found, for instance, in Chapter 7 of this book.

The erbium-doped RFL is a 30 cm-long polarization maintaining structure [43] fabricated by CorActive (peak absorption of 28 dB/m at 1530.0 nm, mode field diameter of 5.7 μm, and 0.25 numerical aperture). A spectrometer with nominal 0.2 nm resolution at 1530.0 nm was employed, with a liquid-N_2-cooled infrared CCD camera as the detector. The pump source was a semiconductor laser operating in the CW regime at the wavelength 1480.0 nm.

A rather long series of RFL intensities $\{I_\alpha(k)\}, \alpha = 1, \ldots, N_s$, with $N_s = 150{,}000$, was generated for an excitation power $P/P_{\text{th}} = 2.92$ well above the RFL threshold where nonlinear effects are relevant, for each one of the 512 wavelengths indexed by k in the interval

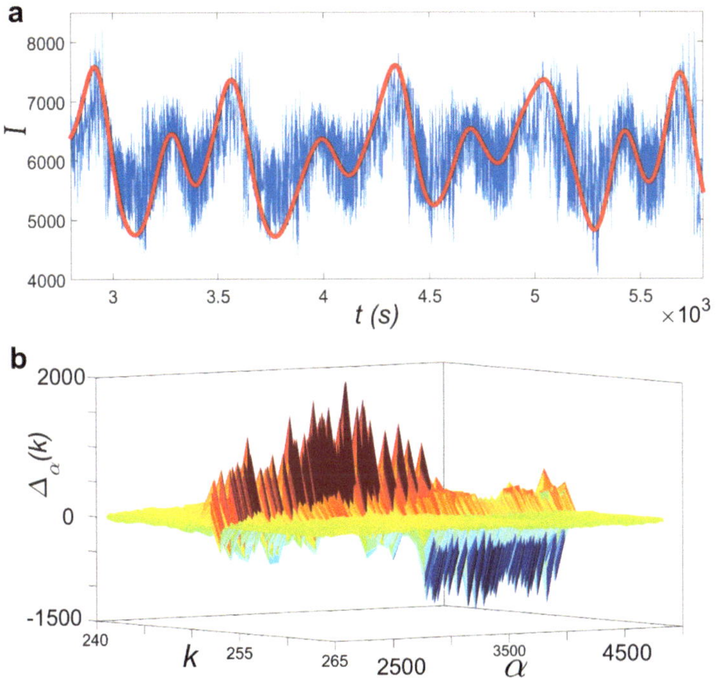

Figure 8.1 Intensity emitted by the erbium-based random fiber laser under an external periodic driving source. (a) Time-dependence of the peak intensity emitted by the erbium-based RFL system pumped by a slowly varying, periodic CW driving source well above threshold (blue line). We show an excerpt of the $N_s = 150{,}000$ RFL spectra. The rapid, small noisy fluctuations are superimposed by a smooth periodic low-frequency signal. The red line is an estimation of this signal using a numerical low-pass filter, with a period $T = 714$ s nearly coinciding with that of the drive. (b) Dynamics of the intensity fluctuations of the RFL spectra as a function of the mode (k) and spectrum (α) labels within one period of the drive, with $t = \alpha t_0$ and $t_0 = 100$ ms. (Reproduced with permission from Ref. [1].)

[1489.0 nm, 1591.4 nm] around the peak emission at 1540.0 nm (corresponding to $k = 254$).

The RFL emitted intensity $I(t, k) \equiv I_\alpha(k)$ for the peak wavelength at $k = 254$ is depicted in blue line in Fig. 8.1(a) as a function of the time $t = \alpha\tau_0$, for an excerpt of the 150,000 spectra. It shows rapid, small noisy fluctuations that are superimposed by a smooth periodic low-frequency signal. We estimated this signal by applying a numerical low-pass filter (red line) to the raw data. We stress that the pump laser intensity also presents long-term oscillations well above threshold with essentially the same period as those of the erbium-based RFL. In order to check this property of the pump laser independently, we rearranged the experimental setup and performed a new measurement at $2P_{\text{th}}$. We found that the period of the pump laser is the same (within $\pm 8\%$) as that of the RFL emitted intensity seen in Fig. 8.1. This oscillatory behavior will be further discussed in relation to Fig. 8.3.

We now investigate the dependence of the photonic spin-glass order parameter overlap distribution $P(q)$ as a function of the time interval $\tau = \beta - \alpha$ (in units of τ_0) between the RFL emission spectra α and β.

Figure 8.2 shows the distribution $P(q)$ for several τ displaying a spin-glass bimodal profile (we notice that the uncorrelated maximum at $q = 0$ is absent), but with a clear asymmetry observed between the maxima around $q = \pm 1$, with strong dependence on τ. This result indicates that subsets of mode configurations separated by a given τ do not correlate (overlap) the same way as subsets separated by different time intervals.

The dynamical asymmetry in Fig. 8.2 contrasts markedly with the *undriven* equilibrium photonic spin-glass phase, in which the side peaks of the bimodal $P(q)$ are *static* and *symmetric*, as discussed.

These findings evidence that a dynamical photonic Floquet phase with spin glass properties has set in the RFL driven by a CW pump source with same oscillation period, which is the dynamic analog of the undriven equilibrium photonic glassy regime. Moreover, the spin glass phase is absent either in a disorder-free fiber without mode localization or under low excitation CW pumping below threshold, when mode localization is absent. Therefore, we conclude that disorder, eigenmode localization, and nonlinearity actually play essential roles to stabilize this photonic Floquet spin-glass phase.

The dynamical signature of the peak asymmetry in Fig. 8.2 can be quantified by defining the asymmetry parameter $\Delta P = P(q \to +1) - P(q \to -1)$ (in practice we consider $q = \pm 0.975$ in the measurements of ΔP). Figure 8.3 plots ΔP versus τ, displaying the same period of the drive and a remarkable agreement between

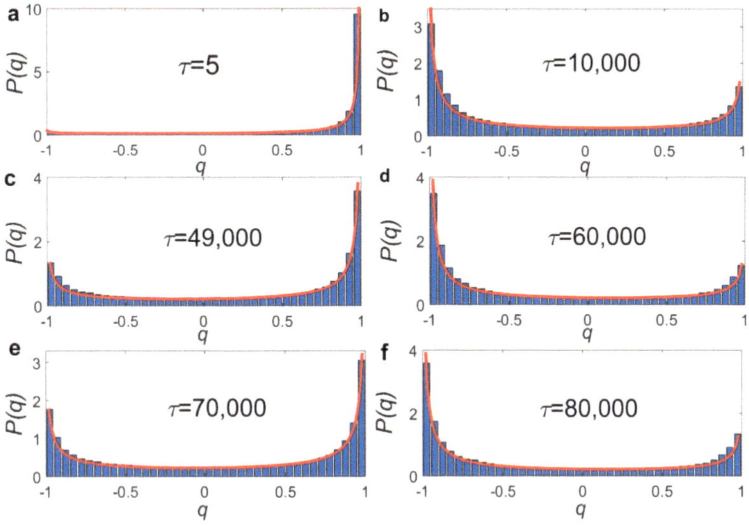

Figure 8.2 Overlap distribution of the dynamical Floquet spin glass phase in the periodically driven erbium-based RFL. Distribution $P(q)$ of correlations between intensity fluctuations of mode configurations separated by a fixed time τ (blue histograms). The spin-glass bimodal profile is evidenced by noticing that the uncorrelated maximum at $q = 0$ is absent. However, a clear asymmetry between the maxima around $q = -1$ and $q = +1$ is observed, with a strong dependence on τ. This contrasts with the static and symmetric side peaks of the bimodal $P(q)$ in the undriven equilibrium photonic spin-glass phase, and indicates that a dynamical photonic Floquet phase with spin glass characteristics has set in the periodically driven erbium-based RFL. In all cases, the red lines show a nice agreement with Eq. (8.13). (Reproduced with permission from Ref. [1].)

experimental data (blue histograms) and the theoretical results (red lines), as discussed below.

First, in order to gain a deeper insight into this dynamical phase, we should initially bridge between the intensity results of Fig. 8.1 and the q-measurements in Figs. 8.2 and 8.3. We start by considering that the driven intensity envelope has a periodic, e.g., sinusoidal, form with k-independent period. In the spin glass phase, an approximate bimodal model for the q-distribution yields $P(q) = p\delta(q-1) + (1-p)\delta(q+1)$, where p $(1-p)$ is the probability of a fully correlated (anticorrelated) overlap with $q = +1$ $(q = -1)$.

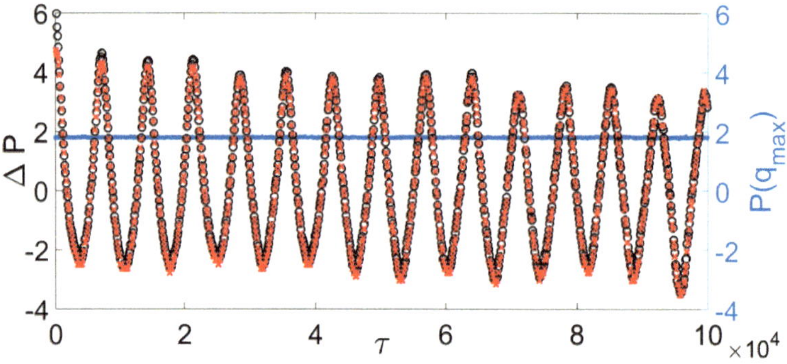

Figure 8.3 Dynamical asymmetry of the spin glass overlap distribution in the periodically driven erbium-based RFL. The asymmetry parameter ΔP is shown as a function of the time interval τ between RFL spectra in the photonic Floquet spin-glass phase above the RFL threshold. A fine agreement between experimental (black circles) and theoretical (red circles) results is noticed. The period is the same of the drive in Fig. 8.1(a) (recall that $t = \alpha t_0$ and $\tau = \beta - \alpha$). In contrast, in the prelasing fluorescent regime below threshold mode localization is absent, no Floquet phase sets in, no asymmetry is found, and the maximum value of the distribution $P(q)$ (blue line) has no dynamics. (Reproduced with permission from Ref. [1].)

Now, since in this case $\Delta P = 2p - 1$ then, by computing the sign function $\Delta I_\alpha \Delta I_\beta / |\Delta I_\alpha \Delta I_\beta|$, we conclude that the asymmetry parameter ΔP related to the spin glass parameter $q_{\max}$ is given by a periodic function with the same period as the drive, in agreement with Fig. 8.3.

Moreover, we can also explain the form of the dynamical asymmetric distributions $P(q)$ in Fig. 8.2 as follows. The overlap parameter $q_\alpha(\tau)$, Eq. (8.11) with $\tau = \beta - \alpha$, is a Pearson coefficient that measures the correlation between the intensity fluctuations $\Delta_\alpha(k)$ and $\Delta_\beta(k)$, with $k = 1, \ldots, N$. Due to the central limit theorem, the distribution of values q_α for fixed τ approaches a Gaussian as N increases, with a convergence rate that actually depends on the specific form of the distributions of the quantities $\Delta_\alpha(k)$ and $\Delta_\beta(k)$. Nevertheless, a much faster convergence to a Gaussian distribution is obtained in terms of a new variable defined by the Fisher transform [53] $z_\alpha(\tau) = 2\,\mathrm{arctanh}[q_\alpha(\tau)]$. As z_α is Gaussian distributed, with mean μ and variance σ^2, the inverse Fisher transform implies the following distribution $P(q)$

with support on the interval $(-1, 1)$:

$$P(q) = \frac{1}{\sqrt{2\pi\sigma^2}} \frac{2}{1-q^2} \exp\left\{ -\frac{[2\text{arctanh}(q) - \mu]^2}{2\sigma^2} \right\}. \qquad (8.13)$$

In the expression above, we observe that μ and σ act as location and scale parameters, setting, respectively, the asymmetry and shape of the distribution $P(q)$. In particular, bimodal distributions related to the photonic spin glass regime are typically obtained for values $\sigma \gtrsim 1.5$. In contrast, unimodal distributions associated with the uncorrelated prelasing fluorescent regime essentially present $\sigma \lesssim 1.5$.

We find that all distributions in Fig. 8.2 are nicely described by Eq. (8.13), with $\sigma = 3.5$ obtained for any τ and $\mu(\tau)$ showing an oscillatory behavior with the same period of the drive. Indeed, by plotting $\Delta P = P(q = +0.975) - P(q = -0.975)$ from Eq. (8.13) as red circles in Fig. 8.3, after a detrend analysis to remove a slight tendency towards negative ΔP values, we observe the remarkable agreement with the experimental data (depicted in black circles), consistently with the photonic Floquet spin-glass phase dynamics set in the erbium-based RFL.

8.4 Concluding Remarks

We end up this chapter by remarking that the above results constituted the first evidence of a dynamical Floquet spin glass phase in a laser-based photonic system [1]. Disorder, eigenmode localization, and nonlinearity are essential ingredients to stabilize the Floquet spin-glass phase in this erbium-based RFL system. The quenched disorder responsible for the spatial eigenmode localization was provided by a remarkable amount of specially designed random Bragg gratings inscribed in the fiber core and correlations among a quite extensive number of output spectra were analyzed. The Parisi overlap parameter and asymmetry properties of the spin-glass order parameter distribution function in the dynamical photonic Floquet phase were nicely described by theoretical arguments.

The experimental validation of the basic theoretically proposed mechanisms for stabilizing phases in driven many-body systems is hard to be overemphasized. In fact, the Floquet mechanism is essential to understand the key role played by the localization effect in the emergence of the spin-glass phase in our experiment. Lastly, we thus believe that our experimental observation of a photonic Floquet phase is a significant step forward in the ongoing collective effort to understand how Floquet phases emerge in actual physical systems.

Acknowledgements

The authors would like to acknowledge the financial support from the Brazilian agencies Conselho Nacional de Desenvolvimento Científico e Tecnológico (CNPq), Coordenação de Aperfeiçoamento de Pessoal de Nível Superior (CAPES), Fundação de Amparo à Ciência e Tecnologia do Estado de Pernambuco (FACEPE), through the projects "Núcleo de Excelência em Nanofotônica e Biofotônica" and "Modelagem de Processos e Fenômenos Físicos em Materiais e Sistemas Complexos," and the Instituto Nacional de Fotônica (INFO).

References

1. Raposo, E. P., González, I. R. R., Macêdo, A. M. S., Lima, B. C., Kashyap, R., Menezes, L. de S., and Gomes, A. S. L. (2019). Evidence of Floquet phase in a photonic system. *Phys. Rev. Lett.*, **122**, 143903.
2. Moessner, R., and Sondhi, S. L. (2017). Equilibration and order in quantum Floquet matter. *Nat. Phys.*, **13**, pp. 424–428.
3. Berdanier, W., Kolodrubetz, M., Parameswaran, S. A., and Vasseur, R. (2018). Floquet quantum criticality. *Proc. Natl. Acad. Sci. USA*, **115**, pp. 9491–9496.
4. Lazarides, A., Das, A., and Moessner, R. (2015). Fate of many-body localization under periodic driving. *Phys. Rev. Lett.*, **115**, 030402.
5. Ponte, P., Papić, Z., Huveneers, F., and Abanin, D. A. (2015). Many-body localization in periodically driven systems. *Phys. Rev. Lett.*, **114**, 140401.

6. Khemani, V., Lazarides, A., Moessner, R., and Sondhi, S. L. (2016). Phase structure of driven quantum systems. *Phys. Rev. Lett.*, **116**, 250401.
7. Else, D. V., Bauer, B., and Nayak, C. (2016). Floquet time crystals. *Phys. Rev. Lett.*, **117**, 090402.
8. Abanin, D., De Roeck, W., and Huveneers, F. (2016). Theory of many-body localization in periodically driven systems. *Ann. Phys. (NY)*, **372**, pp. 1–11.
9. Von Keyserlingk, C. W., and Sondhi, S. L. (2016). Phase structure of one-dimensional interacting Floquet systems. II. Symmetry-broken phases. *Phys. Rev. B*, **93**, 245146.
10. Von Keyserlingk, C. W., Khemani, V., and Sondhi, S. L. (2016). Absolute stability and spatiotemporal long-range order in Floquet systems. *Phys. Rev. B*, **94**, 085112.
11. Yao, N. Y., Potter, A. C., Potirniche, I. D., and Vishwanath, A. (2017). Discrete time crystals: rigidity, criticality, and realizations. *Phys. Rev. Lett.*, **118**, 030401.
12. Lindner, N. H., Refael, G., and Galitski, V. (2011). Floquet topological insulator in semiconductor quantum wells. *Nat. Phys.*, **7**, pp. 490–495.
13. Eckardt, A., Weiss, C., and Holthaus, M. (2005). Superfluid-insulator transition in a periodically driven optical lattice. *Phys. Rev. Lett.*, **95**, 260404.
14. Titum, P., Berg, E., Rudner, M. S., Refael, G., and Lindner, N. H. (2016). Anomalous Floquet-Anderson insulator as a nonadiabatic quantized charge pump. *Phys. Rev. X*, **6**, 021013.
15. Potter, A. C., Morimoto, T., and Vishwanath, A. (2016). Classification of interacting topological Floquet phases in one dimension. *Phys. Rev. X*, **6**, 041001.
16. Hübener, H., Sentef, M. A., De Giovannini, U., Kemper, A. F., and Rubio, A. (2017). Creating stable Floquet–Weyl semimetals by laser-driving of 3D Dirac materials. *Nat. Commun.* **8**, 13940.
17. Potirniche, I.-D., Potter, A. C., Schleier-Smith, M., Vishwanath, A., and Yao, N. Y. (2017). Floquet symmetry-protected topological phases in cold-atom systems. *Phys. Rev. Lett.* **119**, 123601.
18. Berdanier, W., Kolodrubetz, M., Vasseur, R., and Moore, J. E. (2017). Floquet dynamics of boundary-driven systems at criticality. *Phys. Rev. Lett.*, **118**, 260602.
19. Eckardt, A. (2017). *Colloquium:* Atomic quantum gases in periodically driven optical lattices. *Rev. Mod. Phys.*, **89**, 011004.

20. Roy, R., and Harper, F. (2017). Periodic table for Floquet topological insulators. *Phys. Rev. B*, **96**, 155118.
21. Weidinger, S. A., and Knap, M. (2017). Floquet prethermalization and regimes of heating in a periodically driven, interacting quantum system. Sci. Rep. **7**, 45382 (2017).
22. Kyriienko, O., and Sørensen, A. S. (2018). Floquet quantum simulation with superconducting qubits. *Phys. Rev. Appl.*, **9**, 064029.
23. Owerre, S. A. (2018). Floquet Weyl magnons in three-dimensional quantum magnets. *Sci. Rep.*, **8**, 10098.
24. Kalthoff, M. H., Uhrig, G. S., and Freericks, J. K. (2018). Emergence of Floquet behavior for lattice fermions driven by light pulses. *Phys. Rev. B*, **98**, 035138.
25. Zhou, L., and Gong, J. (2018). Recipe for creating an arbitrary number of Floquet chiral edge states. *Phys. Rev. B*, **97**, 245430.
26. Bordia, P., Lüschen, H., Schneider, U., Knap, M., and Bloch, I. (2017). Periodically driving a many-body localized quantum system. *Nat. Phys.*, **13**, pp. 460–464.
27. Zhang, J., Hess, P. W., Kyprianidis, A., Becker, P., Lee, A., Smith, J., Pagano, G., Potirniche, I.-D., Potter, A. C., Vishwanath, A., Yao. N. Y., and Monroe, C. (2017). Observation of a discrete time crystal. *Nature*, **543**, pp. 217–220.
28. Choi, S., Choi, J., Landig, R., Kucsko, G., Zhou, H., Isoya, J., Jelezko, F., Onoda, S., Sumiya, H., Khemani, V., von Keyserlingk, C., Yao, N. Y., Demler, E., and Lukin, M. D. ,(2017). Observation of discrete time-crystalline order in a disordered dipolar many-body system. *Nature*, **543**, pp. 221–225.
29. Ghofraniha, N., Viola, I., Di Maria, F., Barbarella, G., Gigli, G., Leuzzi, L., and Conti, C. (2015). Experimental evidence of replica symmetry breaking in random lasers. *Nat. Commun.*, **6**, 6058.
30. Basak, S., Blanco, A., and López, C. (2016). Large fluctuations at the lasing threshold of solid- and liquid-state dye lasers. *Sci. Rep.*, **6**, 32134.
31. Tommasi, F., Ignesti, E., Lepri, S., and Cavalieri, S. (2016). Robustness of replica symmetry breaking phenomenology in random laser. *Sci. Rep.*, **6**, 37113.
32. Gomes, A. S. L., Raposo, E. P., Moura, A. L., Fewo, S., Pincheira, P. I. R., Jerez, V., Maia, L. J. Q., and de Araújo, C. B. (2016). Observation of Lévy distribution and replica symmetry breaking in random lasers from a single set of measurements. *Sci. Rep.*, **6**, 27987.

33. Pincheira, P. I. R., Silva, A. F., Carreño, S. J. M., Fewo, S. I., Moura, A. L., Raposo, E. P., Gomes, A. S. L., and de Araújo, C. B. (2016). Observation of photonic paramagnetic to spin-glass transition in specially-designed TiO_2 particles-based dye-colloidal random laser. *Opt. Lett.*, **41**, pp. 3459–3462.
34. Gomes, A. S. L., Lima, B. C. , Pincheira, P. I. R., Moura, A. L., Gagné, M., Raposo, E. P., de Araújo, C. B., and Kashyap, R. (2016). Glassy behavior in a one-dimensional continuous-wave erbium-doped random fiber laser. *Phys Rev A*, **94**, 011801(R).
35. Lima, B. C., Gomes, A. S. L., Pincheira, P. I. R., Moura, A. L., Gagné, M., Raposo, E. P., de Araújo, C. B., and Kashyap, R. (2017). Observation of Lévy statistics in one-dimensional erbium-based random fiber laser. *J. Opt. Soc. Am. B*, **34**, pp. 293–299.
36. De Araújo, C. B., Gomes, A. S. L., and Raposo, E. P. (2017). Lévy statistics and glassy behavior of light in random fiber lasers. *Appl. Sci.*, **7**, 644.
37. Turitsyna, E. G., Smirnov, S. V., Sugavanam, S., Tarasov, N., Shu, X., Babin, S. A., Podivilov, E. V., Churkin, D. V., Falkovich, G., and Turitsyn. S. K. (2013). The laminar-turbulent transition in a fibre laser. *Nat. Photon.*, **7**, pp. 783–786.
38. Churkin, D. V., Kolokolov, I. V., Podivilov, E. V., Vatnik, I. D., Nikulin, M. A., Vergeles, S. S., Terekhov, I. S., Lebedev, V. V., Falkovich, G. G., Babin, S. A., and Turitsyn, S. K. (2015). Wave kinetics of random fibre lasers. *Nat. Commun.*, **2**, 6214.
39. González, I. R. R., Lima, B. C., Pincheira, P. I. R., Brum, A. A., Macêdo, A. M. S., Vasconcelos, G. L., Menezes, L. de S., Raposo, E. P., Gomes, A. S. L., and Kashyap, R. (2017). Turbulence hierarchy in a random fibre laser. *Nat. Commun.*, **8**, 15731.
40. Macêdo, A. M. S., González, I. R. R., Raposo, E. P., Menezes, L. de S., and Gomes, A. S. L. (2019). Turbulent intermittency in a random fiber laser. *Atoms*, **7**, 43.
41. Antenucci, F., Conti, C., Crisanti, A., and Leuzzi, L. (2015). General phase diagram of multimodal ordered and disordered lasers in closed and open cavities. *Phys. Rev. Lett.*, **114**, 043901.
42. Antenucci, F., Crisanti, A., and Leuzzi, L. (2015). The glassy random laser: replica symmetry breaking in the intensity fluctuations of emission spectra. *Sci. Rep.*, **5**, 16792.
43. Gagné, M., and Kashyap, R. (2009). Demonstration of a 3 mW threshold Er-doped random fiber laser based on a unique fiber Bragg grating. *Opt. Express*, **17**, pp. 19067–19074.

44. Mézard, M., Parisi, G., and Virasoro, M. A. (1987) *Spin Glass Theory and Beyond* (World Scientific, Singapore).
45. Fisher, K. H., and Hertz, J. A. (1991) *Spin Glasses* (Cambridge University Press, Cambridge).
46. Viviescas, C., and Hackenbroich, G. (2003). Field quantization for open optical cavities. *Phys. Rev. A*, **67**, 013805.
47. Hackenbroich, G. (2007). Statistical theory of multimode random lasers. *J. Phys. A*, **38**, 10537.
48. Jaynes, E., and Cummings, F. W. (1963). Comparison of quantum and semiclassical radiation theories with application to the beam maser. *Proc. IEEE*, **51**, pp. 89–109.
49. Sargent III, M., Scully, M. O., and Lamb, W. E. J. (1978) *Laser Physics* (Addison-Wesley, Reading).
50. Fyodorov, Y., and Sommers, H.-J. (1997). Statistics of resonance poles, phase shifts and time delays in quantum chaotic scattering: Random matrix approach for systems with broken time-reversal invariance. *J. Math. Phys.*, **38**, pp. 1918–1981.
51. Gardiner, C. W., and Zoller, P. (2010) *Quantum Noise* (Springer, Berlin).
52. Cugliandolo, L. F., Grempel, D. R., Lozano, G., Lozza, H., and da Silva Santos, C. A. (2002). Dissipative effects on quantum glassy systems. *Phys. Rev. B*, **66**, 014444.
53. Hotelling, H. (1953). New light on the correlation coefficient and its transform. *J. Roy. Stat. Soc. B*, **15**, pp. 193–232.

Chapter 9

Conclusion and Perspectives

Anderson S. L. Gomes,[a] André L. Moura,[b,c] Cid B. de Araújo,[a] and Ernesto P. Raposo[a]

[a] *Physics Department, Universidade Federal de Pernambuco, Cidade Universitaria, Recife, Pernambuco, Brazil*
[b] *Campus Arapiraca, Universidade Federal de Alagoas, Arapiraca, Alagoas, Brazil*
[c] *Programa de Pós-graduação em Física, Instituto de Física, Universidade Federal de Alagoas, Maceió, Alagoas, Brazil*
anderson.lgomes@ufpe.br

The 21st century is already known by certain key words: It is the century of photonics, when it is contrasted with the 20th century as the century of electronics; it is also known as the century of multidisciplinary work, particularly in sciences, as a development of the individual and sole work form earlier centuries. This book was written within this new reality and exploits photonics as an enabling technology and random lasers (RLs) and random fiber lasers (RFLs) as the tools to implement several new multidisciplinary insights developed by researchers worldwide. RLs and RFLs developments since the 1990s has already found numerous applications in different research areas, including sensing, biosensing, imaging, and optical amplification, among other as reviewed in Refs. [1–13]. The most recent and update review on advancements in RLs and RFLs, including recent applications, is updated until June 2021 [1].

Lévy Statistics and Spin Glass Behavior in Random Lasers
Edited by Anderson S. L. Gomes, André L. Moura, Cid B. de Araújo, and Ernesto P. Raposo

ISBN 978-981-4968-55-3 (Hardcover), 978-1-003-33618-1 (eBook)
www.jennystanford.com

This book deals with yet another multidisciplinary application of RLs and RFLs: their use as platforms for experimental studies, supported by theoretical approaches, to perform studies of several phenomena in statistical physics and complex systems. It started in Chapter 1 with an overview of the several phenomena that were developed in Chapters 3 to 8, besides an introduction to RLs and RFLs. Chapter 2 introduced the role of the feedback mechanisms and modes of RLs, which are of paramount importance for the understanding of some of the analogies that were exploited between photonics and statistical physics. The phenomenon known as Lévy flights, which itself lends to several disciplines [14] and had already been studied within the scope of photonics [15], have found in RLs and RFLs a very interesting fruitful experimental platform, which were the subject of Chapters 3 and 4. Furthermore, and beyond the sole use of Lévy statistics analysis, extreme events and the contributions of higher order optical nonlinearities to intensity fluctuations in RLs and RFLs have been dealt with in Chapter 3.

In Chapters 5 and 6, a traditional subject in statistical mechanics, namely spin-glasses and phase transitions between spin-controlled magnetic states, such as paramagnetic or ferromagnetic states, to the spin-glass phase, has been described in the framework of photonics, using once again the RLs and RFLs as photonic platforms. Indeed, this is one of the most fantastic applications of RLs, since the first observation of replica symmetry breaking (RSB), predicted by Parisi in 1979, as revised in Ref. [16], was first ever demonstrated in a random laser [17], and Chapter 5 brought to the readers a more thorough theoretical background, whereas Chapter 6, written by some of the authors of the pioneering experimental work, described their and other experiments on RSB in RLs and RFLs.

In Chapter 7, RFLs have been exploited as a platform for one of the most intriguing and not completed understood problem in complex system: turbulence. Once again, turbulence is a multidisciplinary phenomenon occurring in several disciplines, as detailed in several review articles and books easily identified by a proper search. Chapter 7 brought to the readers the reviews related to the subject of turbulence-like studies in photonics and in RFLs. Furthermore, the co-existence of turbulence-like phenomenon, the spin-glass phase described in Chapters 5 and 6, and its relation to modes correlations in RLs and RFLs, which have been described very recently, are all included in Chapter 7, making it a very advanced chapter with themes on the frontier of complexity studies in open systems with gain in disordered media. Finally, yet another new experiment was performed using RFLs as a unique platform for photonic demonstration: The so-called Floquet phase, whereby a driven Hamiltonian is employed to demonstrate the effect, was thoroughly reviewed. In fact, as indicated in the perspectives, this is a subject requiring further attention from the scientific community interested in the theme.

It should be reminded to the readers that most of the experimental work described in this book using RLs and RFLs as photonic platforms dates from the year 2004 for Lévy behavior, year 2015 for RSB, year 2017 for turbulence, and year 2019 for Floquet states (see related chapters for references). Several other publications have appeared where the phenomena of Lévy statistics and RSB have been reported, either as further and deeper insights or using novel architectures for RLs or RFLs where the existence of the phenomena is confirmed. These publications are indicated in references [18–37]. We call the readers' attention that no further publication was found in the literature, until the date this book writing was completed, on turbulence-like phenomena and Floquet phase in RLs or RFLs.

From a *perspective* point of view, the use of RLs and RFLs as platforms for experimental work, supported by new or adapted theories to carry on research in statistical physics and complex systems, is a potentially flourishing field.

Regarding intensity fluctuation statistics (from Gaussian below threshold to Lévy at and nearly above threshold, and Gaussian again well above threshold) it is important to understand its behavior in RLs and RFLs with different gain mechanisms, such as Raman gain, as shown in a Raman RFL [22]. Raman RFLs differ markedly from the erbium based RFL discussed in Chapters 3 and 4 because the feedback in the former is provided by the random Rayleigh backscattering in the fiber and the gain mechanism is due to the Raman gain (see Refs. [1, 5]). Moreover, the Raman gain has a much faster (fs) response time than other gain media employed in earlier works on Lévy statistics in RLs and RFLs (ns in dyes and even larger gain response times using rare earth doped ions, such as Nd^{3+} or Er^{3+}, which present, respectively, μs and ms response times. Hybrid RLs or RFLs, consisting of more than one gain media, may also bring novel insights in Lévy-like behavior for the intensity fluctuations. Further work on extreme events, leading to rogue waves in RLs and RFLs should be pursued [38, 39], as well as the role of high order nonlinearities, introduced in Chapter 3.

Replica Symmetry Breaking (RSB) as a signature of photonic paramagnetic to spin-glass transition has been already seen in several RLs and RFLs, and more interestingly, has been studied in the context of simultaneous measurements and with common theoretical background. Further studies on RSB and related phenomena in hybrid RLs or RFLs has already been put forward, as in [21].

Floquet phases in RFLs is still an open and challenging theme for experimentalists and theoreticians. So far, such effect has not been reported in RLs. Furthermore, Floquet phases can be studied from the quantum mechanics point of view [40], which may make the RL/RFL platform again an interesting tool.

References

1. A. S. L. Gomes, A. L. Moura, C. B. de Araújo, and E. P. Raposo. Recent advances and applications of random lasers and random fiber lasers. *Progress in Quantum Electronics* **78** (2021), 100343.
2. F. Luan, B. Gu, A. S. L. Gomes, K.-T. Yong, S. Wen, and P. N. Prasad. Lasing in nanocomposite random media. *Nano Today* **10** (2015), 168–192.
3. L. Sznitko, J. Mysliwiec, and A. Miniewicz. The role of polymers in random lasing. *Journal of Polymer Science Part B: Polymer Physics* **53** (2015), 951–974.
4. S. F. Yu. Electrically pumped random lasers. *Journal of Physics D: Applied Physics* **48** (2015), 483001, 28.
5. D. V. Churkin, S. Sugavanam, I. D. Vatnik, Z. Wang, E. V. Podivilov, S. A. Babin, Y. Rao, and S. K. Turitsyn. Recent advances in fundamentals and applications of random fiber lasers. *Advances in Optics and Photonics* **7** (2015), 516–569.

6. S. K. Turitsyn, S. A. Babin, D. V. Churkin, I. D. Vatnik, M. Nikulin, and E. V. Podivilov. Random distributed feedback fibre lasers. *Physics Reports* **542** (2014), 133–193.
7. D. S. Wiersma. The physics and applications of random lasers. *Nature Physics* **4** (2008), 359–367.
8. H. Cao. Review on latest developments in random lasers with coherent feedback. *Journal of Physics A: Mathematical and General* **38** (2005), 10497–10535.
9. H. Cao. Lasing in random media. *Waves Random Media* **13** (2003), R1–R39.
10. D. S. Wiersma. Disordered photonics. *Nature Photonics* **7** (2013), 188–196.
11. M. A. Noginov. *Solid-State Random Lasers*, Springer, New York, 2005.
12. T. S. Kao, Y.-H. Hong, K.-B. Hong, and T.-C. Lu. Perovskite random lasers: a tunable coherent light source for emerging applications. *Nanotechnology* **32** (2021), 282001.
13. A. S. L. Gomes. Nanocomposite-based random lasers: a review on basics and applications, in *Nanocomposites for Photonic and Electronic Applications*, Elsevier, 2020, pp. 45–79.
14. G. M. Viswanathan, M. G. E. da Luz, E. P. Raposo, and H. E. Stanley, *The Physics of Foraging*, Cambridge University Press, Cambridge, 2011.
15. P. Barthelemy, J. Bertolotti, and D. A. Wiersma. Lévy flight for light. *Nature* **453** (2008), 495–498.
16. M. Mézard, G. Parisi, and M. A. Virasoro. *Spin Glass Theory and Beyond,* 1987, World Scientific, Singapore.
17. N. Ghofraniha, I. Viola, F. Di Maria, G. Barbarella, G. Gigli, L. Leuzzi, and C. Conti, *Nature Communications* **6** (2015), 6058.
18. S. Lepri, C. Trono, and G. Giacomelli, Complex active optical networks as a new laser concept. *Physical Review Letters* **118** (2017), 123901.
19. F. Tommasi, L. Fini, E. Ignesti, S. Lepri, F. Martelli, and S. Cavalieri. Statistical outliers in random laser emission. *Physical Review A* **98** (2018), 053816.
20. A. Tehranchi and R. Kashyap. Theoretical investigations of power fluctuations statistics in Brillouin erbium-doped fiber lasers. *Optics Express* **27** (2019), 37508–37515.
21. A. Tehranchi, V. L. Iezzi, and R. Kashyap. Power fluctuations and random lasing in multiwavelength Brillouin erbium-doped fiber lasers. *Journal of Lightwave Technology* **37** (2019), 4439–4444.

22. J. Li, H. Wu, Z. Wang, S. Lin, C. Lu, E. P. Raposo, A. S. L. Gomes, and Y. Rao. Lévy spectral intensity statistics in a Raman random fiber laser. *Optics Letters* **44** (2019), 2799–2802.
23. F. Tommasi, L. Fini, F. Martelli, and S. Cavalieri. Superdiffusive random laser. *Physical Review A* **99** (2019), 063836.
24. A. R. Hlil, B. C. Lima, J. Thomas, J.-S. Boisvert, H. Iden, Y. Garcia-Puente, L. J. Q. Maia, Y. Ledemi, Y. Messaddeq, A. S. L. Gomes, and R. Kashyap. Rare earth doped PDMS elastomeric random lasers. *Optical Materials* **97** (2019), 109387.
25. E. G. Rocha, E. P. Santos, B. J. dos Santos, S. S. de Albuquerque, P. I. R. Pincheira, C. Argolo, and A. L. Moura. Lévy flights for light in ordered lasers. *Physical Review A* **101** (2020), 023820.
26. S. Bhattacherjee, S. Mujumdar, and S. Ghosh. Lévy flight of superdiffusive light in coupled waveguide lattices. *Physical Review Research* **2** (2020), 012037.
27. A. Sarkar, B. N. S. Bhaktha, and J. Andreasen. Replica symmetry breaking in a weakly scattering optofluidic random laser. *Scientific Reports* **10** (2020), 2628.
28. B. C. Lima, P. Tovar, and J. P. von der Weid. Generalized extreme-value distribution of maximum intensities and Lévy-like behavior in an SOA-based random feedback laser emission. *Journal of the Optical Society of America B, Optical Physics* **37** (2020), 987–992.
29. D. Pierangeli, A. Tavani, F. Di Mei, A. J. Agranat, C. Conti, and E. DelRe. Observation of replica symmetry breaking in disordered nonlinear wave propagation. *Nature Communications* **8** (2017), 1501.
30. I. R. R. González, E. P. Raposo, A. M. S. Macêdo, L. S. de Menezes, and A. S. L. Gomes. Coexistence of turbulence-like and glassy behaviours in a photonic system. *Scientific Reports* **8** (2017), 17046.
31. A. R. Hlil, B. C. Lima, J. Thomas, J.-S. Boisvert, H. Iden, Y. Garcia-Puente, L. J. Q. Maia, Y. Ledemi, Y. Messaddeq, A. S. L. Gomes, and R. Kashyap. Rare earth doped PDMS elastomeric random lasers. *Optical Materials* **97** (2017), 109387.
32. A. Sarkar, B. N. S. Bhaktha, and J. Andreasen. Replica symmetry breaking in a weakly scattering optofluidic random laser. *Scientific Reports* **10** (2020), 2628.
33. P. I. R. Pincheira, M. L. da Silva Neto, M. Maldonado, C. B. de Araújo, A. M. Jawaid, R. Busch, A. J. Ritter, R. A. Vaia, and A. S. L. Gomes. Monolayer 2D $ZrTe_2$ transition metal dichalcogenide as nanoscatter for random laser action. *Nanoscale* **12** (2020), 15706–15710.

34. P. S. Choubey, A. Sarkar, S. K. Varshney, and S. B. N. Bhaktha. Random laser spectroscopy and replica symmetry breaking phase transitions in a solvent-rich polymer thin film waveguide. *Journal of the Optical Society of America B, Optical Physics* **37** (2020), 2505–2513.
35. L. M. Massaro, S. Gentilini, A. Portone, A. Camposeo, D. Pisignano, C. Conti, and N. Ghofraniha. Heterogeneous random laser with switching activity visualized by replica symmetry breaking maps. *ACS Photonics* **8** (2021), 376–383.
36. Z. Zhou, L. Chen, and X. Bao. High efficiency Brillouin random fiber laser with replica symmetry breaking enabled by random fiber grating. *Optics Express* **29** (2021), 6532–6541.
37. J. Kong, J. He, J. Zhang, J. Ma, K. Xie, J. Chen, L. Hong, and Z. Hu. Replica symmetry breaking in cholesteric liquid crystal bandgap lasing. *Annals of Physics (Berlin)* **533** (2021), 2000328.
38. J. Xu, J. Wu, J. Ye, J. Song, B. Yao, H. Zhang, J. Leng, W. Zhang, P. Zhou, and Y. Rao. Optical rogue wave in random fiber laser. *Photonics Research* **8** (2020), 1–7.
39. Y. Song, Z. Wang, C. Wang, K. Panajotov, and H. Zhang. Recent progress on optical rogue waves in fiber lasers: status, challenges, and perspectives. *Advanced Photonics* **2**(2) (2020), 024001.
40. L. Zhou and Q. Du. Floquet dynamical quantum phase transitions in periodically quenched systems. *Journal of Physics: Condensed Matter* **33** (2021), 345403.

Index